DOMINION

NILES ELDREDGE

DOMINION

UNIVERSITY OF CALIFORNIA PRESS
Berkeley · Los Angeles · London

University of California Press
Berkeley and Los Angeles, California

University of California Press, Ltd.
London, England

First Paperback Printing 1997

This edition is reprinted by arrangement with
Henry Holt and Company, Inc.

Library of Congress Cataloging-in-Publication Data

Eldredge, Niles.
    Dominion / Niles Eldredge.
        p.   cm.
    Originally published: New York : H. Holt, 1995.
    Includes index.
    ISBN 0-520-20845-5 (alk. paper)
    1. Human evolution.   2. Human ecology.   3. Population.   I. Title.
GN281.4.E43   1997   96-34174
304.2—dc20   CIP

Designed by Betty Lew

Printed in the United States of America

1   2   3   4   5   6   7   8   9

The paper used in this publication meets the minimum requirements of
American National Standard for Information Sciences—Permanence
of Paper for Printed Library Materials, ANSI Z39.48-1984. ∞

CONTENTS

ACKNOWLEDGMENTS

I thank my colleagues Ian Tattersall, of the American Museum of Natural History, and Elisabeth S. Urba, of Yale University, for their erudite conversation, excellent work, and lasting friendship. I am grateful, too, to my editor Jack Macrae, whose persistence and insight has meant a lot to the production of this book.

FOREWORD

Beyond Nature-Nurture

It strikes me as odd that we are still fighting the old nature-nurture debate. Post-Darwinian Victorians wondered if we were more like apes or like angels. In the modern era of molecular genetics, we are constantly told that we are what our genes have made us. But for every genetically inclined naturist, there is a confirmed nurturist who believes that it is strictly our cultural heritage that shapes our lives—and separates us from the beasts of the field.

It's time to drop the either-or dichotomy of nature-nurture and to confront just what this fusion of consciousness, cognition, culture, and biology that we call "human" is really all about. Indeed, there is an urgent need to do so. We face serious consequences of the explosion in human population triggered by the agricultural revolution some 10,000 years ago. I agree with a growing number of observers that the environmental impact wrought by 5.7 billion humans is beginning to rival the destruction seen in mass extinctions. I also see that it is a dangerous myth that we live divorced from the natural world. Destruction of the global system around us threatens to bring disaster down on our own heads.

My thesis is simple. We humans are indeed a unique species—though how culture and biology have fashioned so completely unusual a species as *Homo sapiens* has not yet been fully recognized. Our ecological history over the past several million years reveals a growing influence of culturally mediated approaches to living in the natural world. We have developed from the small-band existence of our ancestral species, who lived ensconced within the confines of local ecosystems. Those early species evolved and disappeared in lockstep with the other species of their local ecosystems as episodes of climatic change altered their landscapes.

As culture began to come to the fore—when tools showed up in the archeological record, and when our ancestors learned to control fire—we broke away from the rhythmic pattern of extinction and evolution that had been the response of all species to episodes of ecosystem disruption. It was culture, not biology, that allowed us to expand the range of habitats we find comfortable. Our more immediate ancestors began the surge out of our African homeland, a job completed by *Homo sapiens* as we ourselves swept out of Africa 100,000 years ago.

Cultural invention lay behind our ability to spread around the globe. But the real break with tradition, the invention that really changed our position in the natural world, was the advent of agriculture some 10,000 years ago. Taking control over production of our own food supply, *we became the first species in the 3.5-billion-year history of life to live outside the confines of the local ecosystem.* That's what took the lid off our modest population numbers, touching off a population explosion that is accelerating as we move into the twenty-first century.

For 10,000 years, all but a remnant handful of hunting-gathering societies have been living outside the normal, local-ecosystem confines of nature. That is why our cultural heritage proclaims us to be something apart from, even over and above, the beasts of the field. But we have now reached the next crucial phase: *We have become the first species on earth to interact as a whole with the global system.* Other species are spread around the globe as thoroughly as we are. Most of them live as near-parasites around human settlements. The common fruit fly, *Drosophila melanogaster*, is typical: though it has spread around the world, ecologically each population depends entirely on productivity of the local environment. These populations are still members of local ecosystems—however distorted those systems may be. A *Drosophila* in New York knows not, nor cares about, a *Drosophila* in Paris. Not so with us. The global connections among the members of our species are simply unprecedented in the history of life. It is this connectedness, this inwardness, that makes us the first truly economic species on the planet.

Stories are important. I am convinced that hunter-gatherers know full well that they are parts of local ecosystems. I am likewise persuaded that the sages who wrote the lines in Genesis telling us that God intended humans to have "dominion . . . over every creeping thing" realized the special, revolutionary status of their own lives: lives recently freed from the confines of local ecosystems, thus lives seemingly independent of the natural world.

We have to get beyond such myths—accurate as they were 10,000 years ago as an assessment of the human condition. We need an updated story, one that acknowledges that we did not so much leave the natural world as redefine our position in it. It is simply not true that our future is independent in any real sense of the future of the rest of the global system. That global system includes all the world's ecosystems, with all their plants, animals, fungi, and microorganisms, many of which are already under dire threat of extermination at our hands. The global system also encompasses the atmosphere, the oceans, rivers, and lakes, the soils and rocks of the earth's crust. It seems to me that we have formed a bad, 10,000-year-old habit of taking such systems for granted, for overexploiting or even trashing them—simply because we could get away with it and it seemed that we were an independent agent in the world. That time is over.

What does the future hold for us? Surely we have not exhausted our store of cultural "technofixes." How long will we be able to just let things go—population, exploitation of natural resources, pollution? What does genetic engineering promise for solving food shortages and medical needs? What does genetic engineering promise for shaping our own evolutionary future? What, for that matter, will that evolutionary future be like if things just run their normal course—assuming that the species survives long enough for evolution to take hold?

We'd all like to know. But intelligent speculation demands a firm foundation—the most accurate, up-to-date account of who we are and how we got this way. We need a better story, one we can use with confidence as future-telling tea leaves.

That's my aim in this book.

DOMINION

1

Human Futures

Nearly everyone looks to the future with some degree of misgiving. Our feeling that something is wrong transcends the usual "things were better in the old days" and even goes beyond simple millennial jitters. The decades-old threat of nuclear holocaust, though somewhat diminished by recent East-West détente, is still with us, underlain by a less focused malaise: worry that the world is running out of resources, that pollution is mounting, and that competition for remaining resources and simple lebensraum among the mounting hordes of human beings will lead to terrible consequences.

These problems are serious. Studies conducted by the Club of Rome, the WorldWatch Institute, and similar organizations are replete with scary statistics of burgeoning population and depleting resources. There is no question that we must face up to these problems—and no doubt in my mind that if we can face the terrific task of confronting these issues immediately, we can survive the immediate future.

Biology has little to say about these immediate problems; they are, instead, preeminently social issues. Even population control, at least in the short term, is a social rather than a biological issue. And though I take a look at these immediate, pressing issues in this chapter, I will have little to say about them as we probe more distantly into the future.

It would be natural for an evolutionary biologist such as myself to focus, instead, on the long-range evolutionary future of our species. And it is fun to do so—especially for me, who disagrees so strongly with standard prognostications. It turns out that one's basic views on how the evolutionary process works—and how it has shaped the evolutionary histories of all species, especially those of our own lineage over the past 4 million years—determine in large measure what one thinks will happen as the evolutionary ages roll by.

It is the midrange ecological future that concerns me most. We have reached our present precarious position as an outcome of an ecological evolutionary course on which our ancestors embarked at least 2.5 million years ago. And our deep evolutionary history—hence our deep evolutionary future—is a story of shifting positions vis-à-vis our approach to the natural world and its component ecosystems. It is this ecological side of the human story that tells us what is happening now and what our remote evolution is likely to be. But we won't have that remote evolutionary future unless we navigate the intermediate-range ecological future of the next several millennia.

FIRST THINGS FIRST:
SURVIVING THE IMMEDIATE FUTURE

What is going to happen as we turn the millennial corner and plunge headlong into the twenty-first century? Paramount in nearly every prognosticator's mind for the short term is simple *survival.* The world is beginning to look bleak to people with varying perspectives and political persuasions.

It is every generation's seemingly irresistible temptation to pronounce the world's imminent demise. To make matters worse, we are indeed approaching the millennium, and all sorts of dire warnings, biblically based and not, will soon swamp us. Rational minds never pay much heed to the old litanies of gloom and doom. Why should we think that it is any different now? Is there anything special we can point to about the world and the human condition that differs all that much from any other point in human history?

Indeed there is. The difference is *population. Human* population. We are now some 5.7 billion strong and climbing. Ten thousand years ago, we forever altered our relation with the world's natural ecosystems when we adopted agriculture and a settled existence. All of us, that is, but the few indigenous peoples still living as hunter-gatherers. They are as threatened—culturally as well as physically—by the explosion in numbers of agriculturally based peoples as surely as are the vast majority of the millions of species of plants, fungi, animals, and microorganisms.

But that's not all. Ten thousand years ago there were between 1 and 5 million people on the planet. There was plenty of room to expand and move, and resources seemed endless. Now we suddenly see that we are a global species. There are other species, to be sure, nearly as widespread as we. Most are human commensals and parasites: the indispensable (though, in certain strains, morbid and even lethal) *Escherichia coli*, the famous intestinal bacterium. Or *Drosophila melanogaster*, beloved fruit fly of genetics labs, which, like Black and Norway rats, have accompanied humans nearly everywhere.

But none of those species is like *Homo sapiens, our* species, when it comes to its relation with other elements of the natural world. All of the world's species—save our own—are broken up into relatively small subdivisions—local populations that are integrated into local ecosystems. Populations of *Drosophila melanogaster* are far more concerned with the (largely human-altered) local ecological conditions they encounter than they are with far-flung populations of other *D. melanogaster*.

Not so with agriculturally based humanity. Through a 2.5-million-year-long process of substituting cultural for purely biological approaches to living, we have turned our attention from local ecosystems as we go about the daily business of living. We are concerned, instead, with ourselves—meaning, in this case, not so much our personal selves but *each other*. We are an incredibly inner-directed species—the first of its kind on the planet. Based on agriculture and extraction of mineral and other forms of natural wealth from the earth, our entire economic behavior hinges on transactions among humans. Hunting—now confined for the most part to marine fisheries and whaling, and to the harvesting of natural timber stands—remains among the very few exceptions.

We have long been social, and our basic approach to living with nature has long been an attempt to distance ourselves, to use our cultural capacities, our technologies, to afford shelter and increase our abilities to exploit natural resources. How does becoming global change a picture that has been in place for 10,000 years—with elements going back at least 2.5 million years?

The answer lies in what Yale historian Paul Kennedy calls the "Malthusian trap." Rev. Thomas Malthus looms large in the annals of evolutionary biology. It was Malthus's *An Essay on the Principle of Population as It Affects the Future Improvement of Society*, first published in 1798, that supplied a key ingredient to both Charles Darwin and Alfred Russell Wallace, codiscoverers of the principle of natural selection.

Malthus pointed out that if left unchecked, human populations will increase exponentially. Darwin and Wallace grasped the principle and saw that it applies to all species, not just humans. In his *On the Origin of Species by Means of Natural Selection* (1859), Darwin uses elephants to drive the point home: If, he supposes, one starts with a pair of elephants, and if one assumes a reproductive rate of six offspring per elephant over a 60-year interval, in the short span of 750 years there would be 19 million elephants. If left unchecked, the world would soon be standing-room-only with elephants! Clearly, they saw, something is holding in check the explosive potential that lies within each species. The world is not wall-to-wall elephants. Nor is it wall-to-wall anything—except some microorganisms and, increasingly, human beings.

But there is one big difference between elephants (and all other species) and humans—when it comes, that is, to figuring Malthusian limits. All species, including our closest ancestors and living relatives, that have so far graced the earth—with the very recent exception of ourselves—live as relatively small populations disseminated into a variety of ecosystems. It is the local ecosystem that sets the upper bounds on how big populations can get, and the total number of individuals within any species is simply the total number of such limited populations. Stepping outside the confines of local ecosystems, as our species began to do 10,000 years ago with the advent of agriculture, released us from these primordial constraints on our numbers.

Most population experts see our numbers stabilizing naturally sometime during the mid–twenty-first century. While estimates vary at when and at how much, 14 billion seems to come up often as the predicted stabilized number of humans on the planet. But why will there be stabilization at any number? Simply because, instead of stepping completely away from the natural world, we have instead redefined our position in it: We have become a global species, one that— uniquely—interacts as a single unit with the rest of the global system.

It is as if our species has become one single gigantic population. So we will have natural limits imposed upon us after all, and the trick to divining our ecological fate boils down to estimating whether we will stabilize before we end up destroying too many of the world's remaining ecosystems. We may well not stabilize before we have fatally compromised the "ecosystems services"—the carbon, nitrogen, and phosphorous cycles, the production of oxygen and other resources on which we shall always depend.

Which gets us back to the Reverend Malthus's original musings. Malthus was concerned solely with human population. Malthus worried about what would happen when, as seemed to him quite likely, population growth would begin to outstrip the agricultural capacity of society. Famine, pestilence, disease, anarchy—and death. His outlook was grim.

But somehow, society—British society in particular—evaded the Malthusian trap. In *Preparing for the Twenty-first Century* (1993), Paul Kennedy tells us why: Migration was still very much a viable option, and the nineteenth century saw waves of emigration triggered by crop failure (as in the Irish potato famine and emigration of the 1830s and 1940s). Then, too, there was a second "Agricultural Revolution" when crop rotation and other innovations vastly improved agricultural productivity. And then there was the Industrial Revolution itself, which, especially after its effects began to trickle down into the working classes, tended to dampen reproductive rates. Industrialized nations in general have shown declines in birthrates—to levels, at any rate, far below those of "developing" nations.

Was Malthus wrong? Certainly not; unforeseen circumstances merely bought the world a little time. The point about our global distribution and our current 5.7-billion population is that we won't so easily evade that Malthusian trap this time around. To begin with, there is no longer any place to run to. Immigrants have been resented and resisted for millennia, but the fresh waves of illegal aliens seeking refuge in the United States—land of "give me your tired, your poor"—and the public outcry against them are only going to increase.

The population problem is exacerbated by the increasing disparity of distribution of wealth. The division between "haves" and "have-nots," which has sharpened markedly in the last decade within the United States, has well-known reverberations in crime and social unrest. The disparity among nation-states works on a grander scale—but much to the same effect. With mass migration no longer the viable option it was in immediate post-Malthus times, the have-nots will indeed increasingly succumb to famine and all its attendant horrors. Recent events in Africa are a mere adumbration of things to come. The only alternative—if things are left to run their predictable course—will be warfare.

There are, of course, optimists who resist the gloomy tidings, most of them economists with an unbridled faith in the development of markets and wealth. Julian Simon, economist at the University of Maryland, has taken delight in tweaking the Malthusian-inspired analysts transfixed by the horrors they see lying just around the corner. The future will be better than ever, Simon predicts. Population is not a threat, since it is humans themselves who create wealth. The more people born the better, since chances increase that brilliant, creative ones will be born: People who will lead us away from the brink; people who will find ever more ways to avoid the Malthusian trap.

And that *is* the way most of us who devoutly wish that things were *not* going to hell in a handbasket (and that means most humans) traditionally see us avoiding the Malthusian trap. Energy crisis? No problem, "they" (meaning scientists and their funding politicians) will find an alternative source. Except after the bouts with gasoline shortages in the 1970s evaporated, it was back more or less to business as usual—with little in the way of concerted governmental policy or industry-supported research and development to address the problem. Energy research once more became, if not exclusively, at least in large measure the domain of the individual visionary.

Kennedy sees little hope this time that technology will provide the same sort of respite from the Malthusian trap that it provided two hundred years ago. I agree. Unlike the earlier days of the Industrial Revolution when new technologies were developed precisely in those societies where famine loomed, today's advances are coming in the developed nations—and not in the emerging nations of the "third world" where the problem is most acute.

Then there is the problem of resources—*natural* resources. Ecologist Paul Ehrlich, a longtime champion of the environment and advocate of stabilization of human population, locked horns with economist Julian Simon over the immediate future. Putting their money where their mouths were, Ehrlich and Simon made a bet, one which at first glance seemed rather bizarre. Ehrlich bet that the price of copper would rise on the commodities exchange over a five-year period, while Simon predicted the price would actually drop. Simon won the bet.

Beyond the wager lay deeply conflicting views—not only of the short-term human future but of what drives the human social system in the first place. Ehrlich, as an ecologist, knows all too well that resources ultimately dictate population size. There is a ceiling on numbers of individuals within a given population that can be sustained by available resources. Malthus all over again.

Simon, for his part, is keenly aware that human ingenuity time and again overcomes apparent obstacles. Anthropologist Marvin Harris, in his *Cannibals and Kings*, makes a convincing case that humans have, time and again, confronted "energy crises" and won—with a major cultural innovation that not only pushed back the Malthusian trap each time, but in several instances directly contributed to an immediate increase in human population since the "carrying capacity" of the environment was greatly expanded at each event. Thus the initial Agricultural Revolution 10,000 years ago enabled a settled existence, a more or less reliable food supply—and populations immediately expanded as a result. Ehrlich lost his bet because he agreed to measure the effect of diminishing, nonrenewable resources—which is what mining of metallic ores amounts to for humans on this planet—by consulting prices on the commodity exchange. We are indeed running out of precious metals, and so many other natural commodities. But we are far from running out of clever ways to increase efficiency in locating and exploiting these diminishing resources.

But was Ehrlich really wrong? Of course not. He was merely outfoxed. There is no question that we are as limited by resources as any other species on earth. Malthus was right. But what are the limits? Where does the ever diminishing supply of energy resources— nourishment for our bodies, fuel for our engines, plus the matériel for the production of goods—catch up with us?

The specter of running out of essentials first caught up with me in the early 1960s, when Columbia University geologist Rhodes Fairbridge matter-of-factly announced that the world would run out of recoverable petroleum by the mid-1990s. He based his estimate on simple measures of the volumes of the world's great sedimentary basins, and the amount of known reserves in well-explored regions. Fairbridge, it turns out, was a bit too pessimistic: Again, increased efficiency in both exploration and exploitation tactics have staved off the day. But at the present rate of consumption, it can't be too long before we *do* run out of oil. When will that be? The latest prediction is that, with no increase over present rates of consumption, we shall have exhausted known petroleum reserves by the end of the twenty-first century.

Virtually every item in the natural resources inventory is similarly limited. We are cutting down our old-growth forests, in the northern, temperate climes as well as in the tropical rain forests. In principle they are renewable—but not unless we let nature take its course, leaving substantial tracts untouched and allowing cutover areas the centuries or possibly even the millennia they will need to recover completely. There is no sign that humans are willing to do that.

Fisheries and whaling, the last vestiges of substantial food production directly from natural ecosystems, are notoriously depleted. They, too, are in principle renewable, and moratoria on whaling and taking certain species of fish have indeed occasioned the replenishment of harvestable stocks. But the overall outlook is bleak.

Most biological resources are renewable. The problem with their impending exhaustion is one of social planning: regulating harvesting to match renewed growth—an aspect of "sustainable development." Even here, there is a concept of equilibrium—stability in rate of harvest that implies, ultimately, stability in demand—which further implies, if it does not absolutely require in all cases, stability in human population growth. Renewability of biological resources cannot go on in perpetuity in the face of unrestrained population growth. We will use up the resources—and then there will be none.

Oil, coal, and natural gas are an important exception, for they take millions of years to form and accumulate into exploitable deposits. Nonorganic resources, like metallic ores, are another issue entirely. They are nonrenewable for the most part for the rest of the history of the earth. And then there is an intermediate class of resources: inorganic compounds—many vital to life—that are produced by living organisms. Free oxygen, vital to all living things but a few species of bacteria, is replenished daily by the photosynthetic plants and microorganisms of this planet.

Exhaustion of resources need not, in each instance, spell doom for *Homo sapiens*. We are adept at finding alternative resources, of getting by without accustomed supplies. But the loss of each resource drives another nail into the coffin. The question, once again, is: How long will it take? How much can we lose before humans themselves become severely affected? If we ourselves don't put on the brakes to our expanding population soon, it will be done for us—through the exhaustion of resources, the disparity of their distribution in the world, and the social unrest that will surely follow.

How long before disparity in distribution and loss of supply of resources takes its toll? The answer is a few decades, if things persist the way they have been going—and especially so long as our global population continues to expand at its present rate. But maybe not. Maybe there are a few centuries—depending, once again, on what technological fixes can be applied to once again stave off the Malthusian trap.

The point is that our short-term future lies squarely in the arena of human social interaction. Can we control the population? Can we avoid the inevitable disastrous consequences of overpopulation plus national and international disparities in the distribution of wealth? Are there indeed new technological fixes? And can humans live in an increasingly overcrowded and degraded physical environment? These are not so much biological as profoundly *social* issues. Biology comes in when people starve to death—or kill each other off—and there only at the most trivial level: the actual existence of single individuals. It is social decisions that will decide, and for the most part control, the dynamics of the human future in the next few centuries. Malthus applies, but through a complex filter of human social interaction.

I think it 's important to recognize this distinction between the immediate and longer-term human futures. The immediate future is overwhelmingly internalist, dependent as it is on human thought and social action. The next few centuries are sure to be rough. For the rest of this book, I am going to make the assumption that our species will pull through them. And we will do so with a population of some

10 billion with no horrendous cutback in between. But if we are to do so, we absolutely must get a grip on our out-of-control population growth.

Population lies at the heart of our ecological future as well. This second, midrange aspect of our future—already growing in importance and absolutely critical if there are to be successful succeeding centuries and millennia for our species—hinges on our role in nature. We are not as independent, and certainly not as self-sufficient, as we seem to like to think we are. We have, as I have already remarked, altered our position in nature. But we haven't really left it.

The environmental movement is gathering strength. There are immediate concerns, for we ourselves are, largely unwittingly, driving thousands of species extinct each year—as many as 27,000 species (three an hour!), according to Harvard biologist E. O. Wilson. Species are reduced in numbers, and driven over the brink of extinction, largely as an accidental side effect of human utilization of the environment: Agriculture, forestry, and industrial fallout destroy whole ecosystems and poison the atmospheres and oceans.

Human inner-directedness has made us slow to pick up on these concerns. Indeed, the very same Julian Simon who thinks that more people will actually improve the collective economic life of the body politic disparages the notion that we are indeed causing the next big wave of mass extinction—and, in any case, disputes that it matters at all to future human existence even if it should prove to be true.

Part of the problem here is confusing the immediate inner-directed concerns within our species with our environmental problems. Utilization of resources, to be sure, does link the two issues. But they are not the same: The issue of disparity of consumption of resources has profound implications for our immediate survival as a species—meaning the conflicts that are bound to arise. But it also has implications for the relation between humans and an increasingly degraded environmental system in which we live.

It is decidedly not obvious to all of us, living in the last decade of the twentieth century, that a mass extinction of the majority of nonhuman species would be seriously detrimental to the future of humanity. Our ancestors knew how much they depended upon other species populations within local ecosystems, but we don't live as they did. And we do not depend on local ecosystems—or even on the earthly biosphere, the global ecosystem—in the same way our ancestors did. Those of us who see the degradation of ecosystems and the extinction of species as a direct threat to the human future must make a clear case that the mass extinctions of the geological past are real phenomena, and that we are in the midst of a human-engendered extinction spasm right now. And we must be prepared to detail how this latest extinction spasm will eventually have profound implications for the future of humanity.

As we try to delineate the midrange ecological future of *Homo sapiens*, our job will be to see to what extent our species really does interact with the global biospheric system. We are still very much a part of nature, but our global inner-directedness—the tight network of economic interaction within our species on a global basis—has altered our stance toward the natural world. We now interact with the world as a single economic entity—something no species has done before.

This new stance toward nature means that there is a two-way interaction going on. We can see one way very clearly: We are profoundly affecting the global environment. Everyone agrees on that point. The other direction is less obvious: The global environment—its species, its physics, and its chemistry—has a profound effect on us. The more we harm that environment, the more we put ourselves at risk. I return to these themes in more detail, especially in the final pages of chapter 6.

The dissonance between unregenerate optimists like Julian Simon and environmentalists like Paul Ehrlich stems partly from the different time scales they are looking at, and partly from the fact that our most pressing immediate problems are internal and social. But looming right around the corner—the not-too-distant midrange future—the *ecological* future hinges on the very concerns that environmentalists raise.

Human population really does link the short-term, inner-directed concerns of our immediate future with the slightly longer-term prospects for human existence. If overpopulation and the disparity of distribution of resources pose the most immediate threat of social upheaval, they also lie at the heart of our ecological future. Destruction of ecosystems and extinction of species are direct effects of "development." Ehrlich was right: There really is a fairly simple cycle of more mouths demanding more food; more food produces still more mouths, which demand still more food.

Economists like Julian Simon are quick to point to the decline in birthrates in industrialized nations. Raise the standards of the world's underdeveloped nations, and birthrates will level off. A pipe dream, of course. Industrialized nations consume something like twenty-seven times per capita the amount of resources consumed in third-world nations. As Paul Kennedy points out, there is little real prospect for underdeveloped nations across the board to assume such standards. And environmentalists point out that the twenty-seven-fold disparity in consumption of resources really means that the effective population size of an industrialized nation is twenty-seven times greater than its actual head-count—an equalizing factor when it comes to comparing populations of wealthy with impoverished nations. Industrialized nations are every bit as much a part of the population problem as are the poor ones.

So our midterm ecological future is linked, through population growth, with our short-term problems. Unrestrained population growth will eventually lead to ecological disaster if our own internal wranglings—spawned from the same cause—don't get us first. *How* we put the reins on population is the number-one issue facing us right now. There are no simple solutions, but there are some hopeful signs—particularly, recent studies that have shown that education and direct economic empowerment of women, along with ready availability of birth control information and devices, lead to dramatic drops in birthrates.

Our short-term outlook is for economists and social scientists to analyze. The neglected midrange—our ecological future—gets us into some real biology, and invites us to explore our ecological and evolutionary past to chart our changes as we have altered our stance toward nature. And what if we make it though the next few millennia, curbing our population with a new realization of who we are and how we do fit into the natural order of things? Is long-term evolutionary prognostication merely armchair speculation? What will humans be like 100,000 years from now? One million years? Our deep evolutionary past yields some hints at what lies in store—if we do manage to wriggle through the next few hundred years.

2

Nurturing Nature

Wrangling over human identity goes back to the dimmest reaches of the written word. Genesis tends to take the high road: We are created in God's image and appointed to hold dominion over the beasts of the fields. On the other hand, there is an equally long tradition of seeing only one's local group as "the people"—and one's neighbors as sub-human animals. Nor is this an attitude confined to prehumanist time in the ancient world. Sure, Athenians thought in these terms, but so do modern soldiers when asked to kill on the field of battle.

Demosthenes, the Athenian orator who overcame a speech imped-iment to proclaim his opinions, dubbed humankind a "featherless biped." We differ from chickens, according to Demosthenes's sarcas-tic account, only in that chickens have feathers. And that tells us two things: We have *always* sensed that humans are animals. And, at least as long as we have been recording our thoughts, we have been trying to pull ourselves up by our own cultural bootstraps, to loosen the identity with the animal world, to strive to be something better, higher, and at the very least different. To be more like gods than chickens, in fact.

Anthropologists usually think it wise to listen to their informants—but to take their pronouncements with a grain of salt. And so should we as we listen to ourselves contemplating who we are: We should think about what we are hearing. There are really two issues here, closely related as they may be. One is the ages-old nature-nurture hassle; the other is the role humans play in nature. The two are not the same, but they are related in complex ways.

As we look at ourselves right now, it is obvious that we are a culture-bearing animal. I use the term *culture* (with a small *c*) in the broadest possible sense to embrace all behavior stemming from our sense of self-awareness and ability to think, our so-called cognitive capabilities. The nature-nurture debate, in all its various guises, is really just a formalized, modern version of the question: Are humans apes or angels?

This either-or dichotomy is blatantly false. Is intelligence inherited through one's genes, or environmentally determined? The answer has to be yes. The real question is: How much intelligence is environmentally determined and how much resides in the genes? Or, to take another, rather different contemporary element of nature-nurture: Is homosexuality inherited, or is it, too, a manifestation of sociocultural background and other, nongenetic psychological factors? Recent claims of a genetic component to homosexuality may well be borne out by additional research, but it would be truly astounding were it eventually agreed that homosexuality is 100 percent genetically based. Again, reality seems to be more complex than the either-or, nature-nurture choice.

It is tempting to think of culture purely as an overlay on our animal—primate, specifically ape—heritage. After all, we have descended, with absolutely every other living creature on earth, from a long line of living beings that go back a *minimum* of 3.5 *billion* years (the age of the earliest fossil yet found). These were bacteria, and so, 3.5 billion years ago, it is fair to say that *we* were bacteria. No culture, no brains, no appendages—not even discrete nuclei in our (single) cells to house our precious DNA. But there we were, nonetheless.

True animals came along much later—only after the complex nucleated cells that we share with plants, fungi, amoebas, and a host of other microbes had arisen. That happened a minimum of 1.3 billion years ago. Animal life begins to show up in the fossil record about 700 million (0.7 billion) years ago and explodes with a literal bang at the recently revised date of 530 to 535 million years ago. All the major groups of animals—the phyla (including our own, the Chordata)— were there back in those Cambrian days.

We lived in the seas then. True back-boned animals—fishes of various sorts—appeared sometime later. We sprang from one lineage, the bony fishes, which first appeared some 390 million years back. By then, we had brains and fore- and hind limbs to go along with our backbones. By the time we clambered ashore, becoming the first amphibians with functional air-breathing lungs, we had begun the terrestrial sojourn that has brought us to our current environmental dilemma.

When we terrestrial animals developed shelled eggs with yolk sacs and a developing embryo encased in a watery amniotic sac, we were truly free from the water, to which we no longer needed to return to reproduce. While one branch of these earliest reptiles went on to become snakes, lizards, crocodiles, dinosaurs, and birds, our own group was always "mammal-like." Real mammals first showed up some 210 million years ago—right alongside the earliest dinosaurs. But our chance was not to come for another 150 million years, because we mammals lived furtively in the shadows of dinosaur-dominated ecosystems. Mammals quickly radiated in a great evolutionary spurt beginning shortly after the 65-million-year-ago dinosaur die-off. For the first time, some mammalian species grew large, dwelled in herds, and lived off the foliage, while other mammals became specialized predators, eating their herbivorous kin. Mammals now dominated the terrestrial ecosystems that dinosaurs had so long held dominion over.

And what of ourselves? We primates are an ancient lineage of placental mammals who shared part of the Cretaceous Period with dinosaurs. We were, originally, squirrel sized. And while we shared many similarities with another primitive, generalized mammalian group, the insectivores (which include today's moles, shrews, and hedgehogs), we tended to be ecologically flexible. Many early primate species were omnivorous. We remain omnivorous today, cultural explorations into various shades of vegetarianism notwithstanding.

Over the past 60 million years, many primate species have become specialized. Some colobus monkeys, for example, eat only leaves; even some of our human-like relatives living a scant 2 million years ago were obligate vegetarians. But there has always been a strong streak of omnivory permeating the primate approach to nature—a tradition that we have inherited intact.

Monkeys go back at least 30 million years. We know about the earliest ones especially through the excavations of Elwyn Simons, a paleontologist who has been collecting fossils of early higher primates in northern Egypt, on the outskirts of Lake Fayum, for thirty years. True apes appeared soon after the oldest monkeys: The earliest ape specimens yet found, coming from East Africa's Rift Valley system, are 23 million years old. Far from being lumbering bamboo-munching brutes like our modern lowland gorillas, these earliest of the true apes were rather small. They did not have arms elongated for swinging through the trees (in fact, apes apparently didn't develop such "brachiating" anatomies and behaviors until after we had diverged from them only 5 or 6 million years ago). And they appear to have had that traditional primate predilection for an omnivorous diet. These early apes were generalized, ecologically unspecialized, tailless versions of monkeys.

Behavior evolves as surely as skeletal anatomy. In recent decades, anthropologists and ecologists have been focusing on the behavior of modern primate species, from the primitive lemurs of Madagascar up through the greatest of the great apes. The goal is to learn as much as possible about each living species of primate—most of which are under direct threat of extinction. The motive, though, is to learn about our primate kin so that we may learn more about ourselves. It makes sense to look at modern primates, our collateral kin, to get an idea of what our prehuman lives, especially our *behavioral* lives, might have been like. And, naturally, we expect that apes will be more like us than monkeys—and monkeys should shed more light on our behavioral origins than, say, lemurs.

But, as in all things, we must be wary. Mountain gorillas are bamboo and leaf eaters, and adults grow so large they no longer swing through trees. But swing *their* ancestors surely did, whereas ours did not. Gorillas retain those specialized, anatomically elongated, and otherwise modified brachiating arms, whereas our arms, and those of our immediate fossil ancestors, show no such signs.

But with these caveats, it is possible to look at the apes, and to a lesser degree, monkeys, and discover something about what immediately prehuman behavioral life was like. Not surprisingly, there are signs that we don't have an exclusive lock on all things behaviorally "human," or even, for that matter, truly "cultural."

The base of culture lies in our rational minds. And the base of our ability to think, our cognition, boils down in a very real sense to the fact that we are self-aware. We are self-conscious. We, most probably alone among all 10-million-odd species on earth today, are fully aware of our own existence. We know that we are alive—itself a state difficult to define, but we know what it *isn't*. We are all—often painfully—aware that we will die someday.

But are there no glimmers of cognition among other species? I am persuaded by behaviorist Nicholas Humphries's ruminations on the subject. Much as Humphries says he would like to impute self-consciousness to his dog, he finds that he really cannot. But chimpanzees are another matter. Look at any film of a group of chimpanzees, and you will quickly find you can tell them apart, one from another. And they seem to be able to look at one another with (and this is very difficult to judge!) some degree of both curiosity and understanding.

Self-consciousness evolved like everything else about us. Why? What purpose does it serve? Humphries thinks self-awareness, self-knowledge, provides a handy shortcut, a quick reference point to help each member of a social group figure out what's going on with everyone else. It's an intriguing, plausible suggestion. And it appears that, at least in embryonic form, chimpanzees have some measure of this very quality.

Jane Goodall, of course, has spent years studying chimpanzees in the wild, primarily at Gombe, her research site in Tanzania. Goodall thinks chimps have some elements of culture. Most often cited is the ability of chimps to learn from one another how to break off sticks to make probes, used to pry termites from their nests. Even macaques—not apes, but rather old-world monkeys—have been found to use tools and to learn new behavioral tricks which are then learned by offspring and neighbor alike.

So cognition and culture are foreshadowed—embryonically, at least, present in some of our closest living kin. Score one for the animal side of the nature-nurture wrangle. But we also can see that culture is not some sort of deus ex machina overlay on primate existence—something that developed abruptly. We shall see as the story unfolds that culture and physical evolution went hand in hand in the earliest stages of human evolution. Then, beginning less than 2 million years ago, culture began to dominate over purely physical factors *in our basic ecological approach to nature.*

That, in a nutshell, is my overall take on the nature-nurture controversy. We are animals. But over a series of evolutionary events we began to rely more and more heavily on culture—*learned behavior*—in our ecological strategies. Thus began the tension between our animal and cultural selves. A split appeared and, over time, widened into a gulf between our older animal- and newer culture-based approaches to living in the natural world. Sure, we are animals. We eat to grow and maintain our bodies. We breathe. We reproduce. We share 98.5 percent of our genes with chimpanzees—and some vastly smaller, yet vitally important, segment of our genes with the bacteria that line our intestinal walls. But it is our inventiveness and learned behavior that by now dominate our approach to the very acts of living.

The injunctions to think of ourselves as created in the images of the gods, the moral arguments to forsake our animalness, our "baser instincts," all are simply a recognition that we forsook self-unconsciousness and adopted abstract thinking and symbolic communication as a basic approach to living: Exhortations to do better, to develop further our cultural side. Nothing wrong with that. Culture *is* our living strategy, our ecological approach to life. The only problem is that we *are* animals, that we do, despite superficial appearances and attendant myths to the contrary, still live in the natural world.

Ecologically speaking, culture is an enabler, a facilitator. But culture can be a disruption and a source of deep confusion. For we have come to think that we are no longer a part of nature. And that simply isn't so.

CULTURAL AND BIOLOGICAL EVOLUTION

Evolution is a loaded word. It means, to most people in most contexts, "change through time." Yet in the history of most species most of the time, there is precious little change going on. Species mark time during most of their evolutionary histories. The term *evolution* connotes a sort of inevitability, an automatic mechanism for change, which is hardly the case in the evolutionary history of our own species.

So what is evolution? It is tempting to call it just plain history: what happens to a species, or a language, as it passes through time. But that isn't very satisfactory, either, for then there is the obvious question, What's the difference between the two? Why do we need a separate word, *evolution*, when *history* would do just as well?

While many evolutionary biologists are content simply to work out what happened in evolutionary history, others are preoccupied by the underlying causes of evolution. What drives evolution? Why do species sometimes change—and sometimes remain the same? Ever since Darwin, evolutionary theorists have sought to refine our understanding of how the evolutionary process actually works.

Historians have been much more reluctant to investigate processes of sociocultural change. Anthropologists have taken something of a middle road, flirting with evolutionary ideas since the nineteenth century, but remaining on the whole less preoccupied with causal factors of historical stability and change than their counterparts in biology.

Making sense of causal processes underlying what is, after all, dead and buried history is a tricky proposition. How can a paleontologist hope to study the dynamics of genetic change with the long-dead fossils lying on the laboratory table? The trick lies in seeing that history, in a general way, does seem to repeat itself. Evolution leaves recurrent patterns frozen in the fossil record—patterns that can be explained through what we know about the causes of genetic stability and change in modern organisms. Taking actual evolutionary history into account helps refine our grasp of how the evolutionary process actually works.

A case in point, and one very relevant to human evolutionary history: In the early 1970s, Stephen Jay Gould and I brought back into prominence the long-known but long-overlooked fact that once species first appear, they do not change all that much throughout the remainder of their histories—which might go on for hundreds of thousands, often millions, of years. We called this general pattern *stasis*.

I first encountered stasis when, as an enthusiastic graduate student, I began an ambitious study of the Devonian trilobite *Phacops rana*. Trilobites, extinct relatives of crustaceans and horseshoe crabs, were among the earliest of complex animals to show up in the fossil record—some 535 million years ago. I expected to find abundant evidence of smoothly transitional, typically Darwinian (or gradual) evolutionary change in my trilobites. After all, their lineage lasted some 8 million years (beginning around 385 million years ago). Easy to collect and usually well preserved, occurring in a variety of ancient marine environments spread out over what is now eastern and central North America, the *Phacops rana* lineage *should* show every sign of slow, steady gradational evolutionary change—according, that is, to the prevailing conventional wisdom of the past century. What I found instead was obdurate intransigence, a sort of stony implacability, for all my trilobites, through thick sequences of rock and correspondingly long stretches of time, looked bafflingly alike. It is this sameness, this evolutionary stability, that Gould and I labeled *stasis* as we sought a general explanation for the patterns we were seeing in the fossil record.

Stasis collides utterly with standard evolutionary expectations going back to Darwin himself. The non–historically inclined side of the evolutionary debate relies on pure extrapolation of what it can learn about genetic change in modern populations—in the laboratory or directly in nature. And there is no doubt that such information is important—even vital to a complete understanding of the evolutionary process.

But consider what happens with this pure extrapolationism, unfettered by historical fact. Darwin himself argued, in essence, that just as a little change can accrue over a few generations, we can imagine that great changes might accrue over truly vast amounts of geological time. Darwin was looking at the fruits of domestic animal breeding programs in the mid-nineteenth century (he was especially knowledgeable about the activities of pigeon fanciers). Modern-day evolutionary-minded geneticists are more fond of fruit flies than pigeons. But the logic remains the same: If we can manipulate our populations into changing a bit over a few generations, think how much change could automatically accrue over long periods of time.

The historical pattern of stasis, known in Darwin's day, was dismissed as an artifact of an extremely spotty, incomplete fossil record. Gould and I, and a number of like-minded paleontological colleagues, have managed to convince most of the scientific community that stasis is a real, dominant historical evolutionary pattern—one, their more prominent spokesmen now admit, unanticipated by the pure extrapolationism of traditional evolutionary theory.

History matters! Patterns of history tell us how the various processes of genetics really do fit in with the ecological exigencies of living in the real world—an interaction that leaves behind it a trail of stability and change that *is* evolutionary history. We need continued analysis of the genetics and ecology of living creatures; we need, as well, careful study of historical pattern to see how the processes really go together to yield evolution. And then, and only then, are we in a position to say something cogent about the future. Short term-process + evolutionary pattern = coherent theory of temporal process. Past, present, and future.

Analysts of human history—especially human history since the advent of the written word only some 8,000 years ago—have been somewhat reluctant to confront general processes of history. To do so would require a grasp of a prodigious assortment of short-term socio-cultural dynamic processes—running the gamut from language, the invention and use of specific cultural items (tools in the broadest sense, from axes to computers), on up through large-scale features of socio-cultural organization, such as the nature and fates of nations. Such would have to be integrated with a search for general, repeated patterns in the human historical record—a tall order, much more difficult even than deciphering evolutionary patterns and their probable causes from the fossil record.

Some historians, though, have been looking for patterns of history. My favorite is a somewhat obscure figure in the annals of American historiana: Frederick J. Teggart. In the years immediately following World War I, Teggart pointed out that the newly devised social-studies curriculum for California secondary schools raised many general questions about social science as a whole. Teggart saw that such a curriculum demands a general theory of social structure and change to serve as the theoretical underpinnings for the course of instruction. He looked for one—and was astonished to find nothing available. So he did the next best thing: He constructed one himself.

In his search for preexisting general theories of history, Teggart read Charles Darwin. What he said about Darwin's theory of evolution was remarkable, given Teggart's background in social science and his presumably complete lack of hands-on experience with the fossil record. He said that Darwin's was a theory that paid scant heed to the facts of history! That's precisely what Gould and I were saying in the 1970s. Teggart saw it a full half century earlier, writing with great conviction in his book *Theory of History* (first published in 1925, and available now, combined with his earlier essay *Processes of History*, as *Theory and Processes of History*).

Teggart disdained any general theory of evolutionary change— biological or cultural—that failed to address what he called "events." Teggart meant categories of repeated, similar events—what I have been calling historical patterns. According to Teggart, Darwin's theory comes up wanting because Darwin didn't acknowledge the prevalence of stasis. Just how Teggart tumbled onto the ubiquity of stasis, of *non*change, is by no means clear to me; by the early twentieth century, paleontologists had all but forgotten stasis. Stasis was so out of sync with the extrapolationism of Darwinian thinking that it had become a pure embarrassment, a symbol of the inadequacy of the fossil record for telling us anything really interesting about how life evolves. Teggart's forthright assertion that there was something wrong with Darwin's approach, rather than with the fossil record, was amazing for its time and provenance, coming as it did from a historian of human affairs, rather than a paleontologist.

Turning to his own specialty, Teggart saw three main patterns, signals that crop up over and over again in human history. To begin with, there *is* some change inherent in almost any sort of cultural item one can specify. As time goes by, as each generation inherits the customs and tools of its predecessors, there is inevitably some change. Language is notoriously labile. My teenage sons in New Jersey, for example, routinely use the word *stick* to mean "leave"—confusing my wife and me no end until we finally caught on.

Such change is more like a random drift than a concerted movement for change in any particular direction. There is nothing inherently better or more efficient in using *stick* rather than *leave.* It is more a matter of modish taste, and indeed the motive for its use is group identity, the definition of an in-crowd that understands, and a group of benighted outsiders—especially parents—who don't. As such, *stick* is far more an impediment than an improvement in communication.

Such drift over the generations is hardly the stuff of invention of new languages, or so Teggart thought. The point is crucial, for Darwinians have always argued that small, generation-by-generation change is the stuff of large-scale evolutionary change: All you need is enough time, and change will take care of itself. So too goes the generally unexamined assumption that all large-scale cultural differences between peoples arise through the very same small-scale, generation-by-generation change accrued through time.

Chaucerian English, after all, is quite different from Elizabethan language, which is, in turn, rather different from modern English—in all its various guises. There is a definite historical linearity of change running through the English language over the past 1,000 years—a change that would no doubt make it impossible for any speaker of modern English to converse easily, if at all, with Geoffrey Chaucer or any of his contemporaries.

But is such drift through time really all there is to language evolution? Forgotten in the urge to extrapolate small-scale change into a general theory of language history is a very different component: regional differentiation—a process that leads to rapid linguistic shifts in some instances and extraordinary stability in others. American

English in general has supposedly retained more overt Elizabethan elements than can be found anywhere in the British Isles. And isolated pockets (Anglo-Saxons on Smith and Tangier Islands in Chesapeake Bay; Gullah blacks on the Carolina sea islands) are said to retain strong elements of speech patterns characteristic of the earliest English-speaking settlers.

On the other hand, regional differentiation is a sure path to rapid change. It is as true of animal and plant evolution as it is of cultural innovation. The fossil record tells us that most evolutionary change comes when parts of a species become isolated. Each goes its separate way, developing its own adaptations to its local environmental setting. Human biological evolution was as much dependent on this process of "speciation" as any other group of animals. Closely related species tend to live in adjacent, but geographically disjunct, regions, especially early on after their mutual divergence. The Isthmus of Panama emerged as dry land only some 3 million years ago, splitting up marine species into Pacific and Caribbean populations—and leading to the evolution of a number of new species of mollusks.

Similarly, Europe's language map reveals a series of dialects and separate but closely related languages that replace one another regionally, existing cheek by jowl. They were developed, of course, in pre–Industrial Revolution days, when communication was limited to infrequent physical contact.

But, Teggart felt, while some sort of drifting change permeates virtually all cultural items through time, by far the stronger signal is stability. Culture is often as intransigently resistant to change as species themselves seem to be.

Cultural stability is indeed a strong signal in human history, going back as far as 2.5 million years when the very earliest stone tools show up in the archaeological record. My favorite example from the era of recorded history (the post–Agricultural Revolution, with its emergent nation-states and written record) lies in the cultural kit bag of Egyptian peasants. For example, Egyptian tombs record two basic ways of lifting up water from the Nile and its canals to irrigate the fields. One, the shadouf (its modern Arabic name), is a clay pot that swings from a long pole, counterbalanced by a weight on the other end—a lever system that enables a relatively effortless, if slow, means of getting the job done. Some 5,000 years ago the waterwheel appeared, an idea perhaps borrowed from the Persians. The waterwheel is a horizontal, spoked affair (usually powered by an ox walking in an endless circle) that is hooked, through a system of gears, to a vertically rotating circle of clay pots that dip into a well and pour out their contents as they come swinging up and over. A third method is also ancient: the Archimedes screw (tambour)—literally a large screw inside a cylinder, used to pull water up an incline to the field above.

In 1979, I saw the Egyptian landscape dotted with working examples of all three mechanisms (though the main working waterwheel I saw doubled as a payment-expected tourist photo op). In contrast, in 1990, the one shadouf that I saw was a lonely tourist specimen—standing idly beside a Japanese-manufactured gasoline pump that was spewing out water many times faster than the ancient methods ever could. Here was 5,000 years suddenly displaced by cultural innovation. And that raises the question, If stability is the main signal of most of cultural history, swamping out what little drifting, aggregate change might accrue, whence, then, real innovation? When, how, and under what circumstances do we see invention—true cultural change—take place? That is, after all, what most of us still think of as "evolution": change, the invention of the truly new.

Innovative change was Teggart's third and final category of historical pattern. He tended to equate change with real events. Change is often sudden. Far from being the outcome of a long, slow, steady growth, change most often appears abruptly. This was the second element of Teggart's thesis that grabbed my attention when I first stumbled on his writings in the late 1970s. For when Gould and I resurrected stasis as the main signal in the history of species, we also pointed to the corollary observation: When evolutionary change does occur, it tends to happen relatively quickly. Biological evolutionary change seems almost sudden when compared with the vastly longer periods of time when nothing much seems to be happening to species.

As a source for cultural innovation, Teggart was especially intrigued by collisions between cultures. For example, Egyptians got the wheel when they were invaded by the Hyksos, who showed up in their horse-drawn chariots during the interregnum between the Middle and New Kingdoms. Likewise, the introduction of the powered Japanese water pump was a (considerably more peaceful) invasion, an event that introduced cultural innovation to a land where water-lifting devices had remained the same for millennia.

But we cannot always say that change in human cultural history is always a matter of collision between societies; change cannot always come from elsewhere, for that simply begs the question of where the innovation was first developed. Where did the Hyksos get those horse-drawn chariots? *Somebody* had to have invented the wheel—even if we know that the Egyptians simply adopted it from their temporary conquerors. How does the process of cultural innovation and subsequent stability really work?

Evolutionary Process:
Adaptation in Cultural and Biological Evolution

Patterns of cultural evolution, then, often look very much like evolutionary patterns. In both realms, stability is a strong historical signal.

True change—not the minor drifting that goes on through time, but real, substantive change—seems much more rare. It tends to show up rather suddenly. And in both systems, change seems to come through a process of differentiation, where populations or cultures tend to go their separate ways when isolated one from another.

But similarity of patterns is not the whole story and by no means implies that the mechanisms of cultural and biological change are identical. The heart of the biological evolutionary process is natural selection, whereas cultural evolutionists speak of "cultural selection." There are some valid analogies between them, but there remains a fundamental difference between the cultural and biological realms, a difference that was critical as ancestral human species shifted from a biological to a cultural ecological strategy over the past 2.5 million years.

Cultural and biological (natural) selection are simply not the same thing. Both produce patterns of stability and change, and historical patterns of cultural evolution often seem to parallel those we see in the evolutionary history of life. But the two do not operate the same way. And, with the exception of only the dimmest, earliest reaches of human prehistory, the two do not work in concert. Cultural and biological evolution are decoupled—in principle and in reality. We need to go back to Darwin's original formulation of natural selection to see that this is so.

Darwin's lasting gift to us was the certainty that life has evolved. He also pinned down the fundamental mechanism governing stability and change of the anatomical features of plants and animals, including our own corporeal selves.

That mechanism is natural selection. Natural selection remains as valid an explanation of why organisms seem to fit their surroundings so well as when Darwin first succinctly and convincingly propounded it—and when Thomas Henry Huxley remarked not long after how extremely stupid it was not to have thought of it himself.

The gist of natural selection is simple enough. Darwin realized that populations of all species are limited; by no means can all offspring born each generation possibly survive and reproduce. We know this is so simply because no species—with the current conspicuous exception of our own—is engaged in the sort of runaway expansion that would automatically happen if all young were indeed able to grow up and reproduce. Even in our own case, infant mortality remains a conspicuous element of the human condition. Recall Darwin's hypothetical example of exploding elephant population growth if left unchecked by resource limitations, disease, and other environmental conditions. For most species, resources—meaning, for animals, the availability of food, water, and nutrients—put a ceiling on the number of individuals of each given species that can thrive and reproduce.

Only those individuals that best meet the tough requirements of eking out a living will, on the average, reproduce. And when organisms do reproduce, they pass their genes along to the next generation—100 percent when reproduction is asexual (uniparental) and 50 percent when reproduction is sexual—as, of course, it is for humans. And those genes are the recipe for success for their children. For if conditions remain substantially as they were when their parents were competing, relatively successfully, for those resources, their children will tend to be successful as well—since they will have inherited the features that proved to be superior for making a living in their parents' generation. If conditions don't change all that much, what worked for the parents will work for their offspring—and the status quo will remain.

Darwin saw that there is typically a great deal of genetic variation in each population. We now know that mutations are constantly occurring, changing isolated bits of genetic information here and there within each individual organism. Many mutations are harmful; many make no real difference one way or another to an organism's life. Others actually seem to help.

Part and parcel of our sexual reproduction is a reshuffling of genes as sperm and eggs are manufactured. The one-half of my genes carried by my older son are different—to some indeterminate degree—from the one-half of my genes carried by my younger son. Each child (with the sole exception of identical twins) sports a different spectrum of parental genetic information. Variation is rampant in all sexually re-producing populations, and it is this heritable variation that presents the opportunity for evolutionary change.

And that means that when times *do* change, what worked best in previous generations may no longer do as well as some other set of characteristics. Other traits may now confer greater success than those that worked in the past, and these different traits begin to show up in greater frequencies in succeeding generations. Evolutionary change—change in the distribution of genetically heritable features—begins to accrue.

The celebrated case of "industrial melanism" in the British popula-tions of peppered moths remains an especially instructive example. The moths occur in two distinct color phases—one light, the other much darker. When clinging to the trunks of trees, the moths blend in with their background, their cryptic coloration making it difficult for bird predators to spot them. When the Industrial Revolution, with its belching smokestacks, began to darken England's trees, the darker peppered moths became more abundant over the yearly generations—for their color better matched the majority of trees. Efforts to clean up the atmosphere, however, turned the tables, and natural selection increased the frequency of lighter moths as England's trees lightened up. I should also point out that this example, along with many other well-established cases of natural selection in the wild, highlights the essential stability inherent in many patterns of long-term evolutionary change: Toggling back and forth between dominance of one color form over another (mediated as that change is by natural selection), a sort of oscillating stability—rather than long-term net directional change—is the natural result.

What, if anything, is there on the *cultural* side that resembles natural selection? If the general patterns of biological and cultural history are so similar, do they spring from the same root cause? What marshals evolutionary stasis and change in human cultural history?

Humans, for all our folly, are pragmatists. True, we rush after fads, embracing the latest craze from dance steps to kitchen accessories, only to drop them when something new and different comes along. My wife gave away an awful lot of extra fondue pots after we got married in 1964, when the twist was all the rage on the dance floor. But I am writing—composing and typing—this book on a personal computer, using a word-processing system that checks my spelling, allows me to alter the format of the type, and, best of all, makes correcting and retyping a breeze.

New inventions often displace the old—and the analogy with natural selection seems clear. Biological and cultural evolution definitely are biased toward the better mousetrap: the new, improved version of an existing structure, or cultural artifact, that simply works better than the preexisting version. Cultural historians and anthropologists have explored this analogy intermittently for years. Some call the cultural version of selection just that, "cultural selection." Others come right out and maintain that there is no real difference between the two, that what some choose to call "cultural" selection is just another form of "natural" selection. After all, biologists don't in any real sense own natural selection—or so goes the argument.

But there are fundamental differences between the cultural and biological domains of selection—whatever we choose to call them. Both stem from the obvious fact that culture is a product of conscious, even rational, thought. Natural selection works on the heritable differences between individuals, on the variations present in each generation handed down genetically from parents to offspring. Not so with selection of cultural items. Invention of word processors for home computers was not picked up and passed along through the genes. True, kids growing up in homes with computers already in place are more likely to pick up the habit of using them (for games first, homework later!). But that has nothing to do with genes and everything to do with direct introduction—learned experience coming years after those kids had come into their genetic inheritance.

In other words, it's not just the biological children of the clever souls who scaled down computational devices—making computers both small and inexpensive enough to invade the home market—who in fact have all those millions of PCs right now. In biological evolution, new traits increase in frequency in a population if and only if those that have those genetic characteristics in fact thrive and outreproduce members of their generation who lack the traits. But cultural systems are different: We humans learn—fast—from our fellows; we ape each other relentlessly.

Darwin wrote *On the Origin of Species* nearly a half century before biologists began to figure out how the process of heredity actually works. His own ideas on the subject were far off the mark. All he really knew (and all he really needed to know to formulate the notion of natural selection) was simply that organisms do, in fact, resemble their parents, and that all that variation out there in wild populations is, for the most part, heritable. But Darwin was a pluralist, and in later editions of the *Origin* (the sixth edition is the most widely available today), succumbing to critics, he widened his argument to embrace additional possible mechanisms of evolutionary change beyond pure natural selection. One of them was the "inheritance of acquired characters"—so-called Lamarckian inheritance.

Early forerunners of Darwin—biologists who had posited evolution in one way or another—had from time to time suggested that change could come about if organisms acquired a feature through constant use. The new feature, it was supposed, could then be passed along to the next generation. Giraffe ancestors, in the classic tale, would stretch their necks more and more to reach the leaves at the very tops of trees; modern giraffes got those hyperextended necks through a long series of generations of trying to reach beyond the limits. It took August Weismann, a German biologist active in the latter part of the nineteenth century—well after 1859—to show that inheritance is caused by factors in the sex cells, sperm and eggs, and that what happened to an organism's body during its lifetime could not possibly affect the features inherited by its children. Weismann's dictum has stood the test of time well, despite occasional challenges that still crop up periodically. Inheritance of acquired characteristics is out as a viable theory of inheritance. Biological evolution cannot possibly proceed in such a fashion.

But computers are acquired. And kids learn to use them at school even if their parents don't own one. In natural selection, a trait must directly affect its bearers' reproductive success in order to spread. Crucially, a cultural trait can be picked up and spread widely with no regard to the reproductive proclivities of those who pick up and transmit the trait. Computers may well enhance economic efficiency in the lives of those who adopt them over quill pens and electric typewriters, but this is so whether the writer has kids or not, or whether whatever kids she might have had came along before or after the computer ended up in the home office. Learning is a form of

Lamarckian inheritance—and a means of much faster spread as new traits are adopted by millions virtually overnight.

Nor is that all. Kindred to debunked notions of Lamarckian inheritance of acquired characters in biology is the notion that new features will occasionally spring up to meet the needs of organisms directly. New genetically based traits—mutations—show up in every generation. When they were first discovered, mutations seemed invariably harmful, and many were downright lethal. Early geneticists had a hard time reconciling mutations with Darwinian natural selection, until improved experimental techniques eventually revealed that many mutations have only slight effects (difficult to detect in the early days of genetics) and may well be beneficial—or at the very least neutral, with neither ill nor positive effects.

The important point is that mutations do not arise as a spontaneous response to some new needs that an organism might face. Mutations are biochemical events—errors in copying genetic information as cells divide. They always have a cause, whether through a biochemical accident within the cell itself or through some outside agent. Mustard gas and radiation are common mutagens. But here is the crucial point: Whatever the actual physical cause of a mutation, the event itself is random insofar as whatever good or harm it might do the organism in the environment that it is in. Even if the cause of the mutation is environmental, the mutation won't help the organism survive in that environment; radiation does not cause mutations that help an organism resist the deleterious effects of radiation. Radiation causes mutations that, more often than not, are harmful—including the genesis of some cancers.

Biological mutation does not act to push evolution in any particular direction. Natural selection does that, winnowing an array of genetic variation whose ultimate source is mutation. Now consider the human cultural situation again. Humans invent cultural items to serve a purpose; they immediately begin to tinker with them to improve the design. The first automobiles were motorized buckboards—horse-drawn carriages co-opted for the purpose. Cars passed beyond the buckboard-and-carriage stage virtually overnight, as soon as it was realized that an internal combustion engine placed on a four-wheeled conveyance would indeed work.

Thus the other major difference between cultural and biological evolution: We humans always *direct* our variation. Whatever the subtle mix of conscious planning and creative insight and inspiration that yields the truly new in human experience, the result is a product that more or less does the trick in meeting whatever the need is at the time: a way to open a can, a weapon that can steer itself, or just a new song.

Biological evolution meets needs—improves adaptations—only by sorting through each generation's genetically varied heritage. Mutations, its ultimate source of novelty, are never produced in direct response to some sort of perceived need. Human inventiveness, the entire process of human cultural change, works much more quickly and contains a spark of conscious originality that sets it far apart from the mindless, slogging persistence of biological evolution.

Culture has become *the* human adaptive strategy, *the* way of approaching the natural world and wresting a living from it. It is the protective web and the set of cunning devices that enable us collectively to survive, indeed to thrive. It is such an all-enveloping cocoon, such a successful ecological strategy, that we have forgotten, to an astonishing degree, that we came out of a much more intimate relationship with the natural world. And culture has, ironically, blinded us to the fact that we are still very much a part of that world. Our culturally based ecological strategy has worked so well that we no longer see it for what it really is: a means of existing in the natural world.

Instead, we seem to exist in a culturally devised world of our own making. We are preoccupied with ourselves—ourselves as individuals and with each other. We are astonishingly inner directed. Our culturally imbued skills have been so successful that we think we have won— we have sought and gained dominion over nature. We are self-sufficient, or so we think. How, exactly, did we get this way?

3

The Way We Were

Of all the pop-biology books proclaiming the animalness of humanity and the gene-rootedness of all our behavior, none resonated more with the public than Robert Ardrey's *African Genesis* (1961). Ardrey's book led the charge, proclaiming as stuff and nonsense the uniqueness of *Homo sapiens* long before gene-struck evolutionary biologists jumped into the fray. Ardrey was a playwright and screenwriter, giving him two advantages: the ability to produce a lyrical text, and the status of an outsider, often relished in a world where expertise can be, at best, an equivocal virtue.

Ardrey thought that the violence so prevalent in modern society flows directly from our biological heritage. Wars, pogroms, and random killings alike owe their genesis to the behavior of our remote ancestors—our African ancestors. We evolved, or so Ardrey claimed, from a species of "killer ape": *Australopithecus africanus.* Thus the "it's all in our genes" approach to understanding human nature began with an especially dark portrait of our own species—and with a very particular point of view of the ecological behavior of one of our ancestral species. We *are* beset, of course, with wars, pogroms, and random killings. But was it accurate to blame our evolutionary patrimony, our genetic heritage, instead of more proximate factors of relative wealth and the competitions and animosities it engenders? Is it fair to blame our genes, and one particular ancestral species, rather than shouldering the blame ourselves? Is murder ingrained instinct or culturally mediated behavior?

Ardrey's thesis hung on his interpretation of the ecological behavior of this ancient relative of ours, *Australopithecus africanus.* Not unnaturally, Ardrey embraced the views of Raymond Dart, the discoverer of this ancient species. As a young Australian anatomist teaching at the medical school in Johannesburg, South Africa, Dart had become interested in the baboon and other fossils that had been turning up in some of the commercial quarrying operations in limestone deposits that dotted the countryside in the Transvaal. The quarry manager of the limeworks at a place called Taung had taken to sending him odd lots of promising fossils as they were exposed in the course of normal operations.

One day in 1924 (supposedly while he was dressed in formal garb awaiting the start of a wedding ceremony—these stories get better with age) Dart opened a shipment from Taung and saw the exposed braincast of a small apish creature. Dart was a neuroanatomist and saw immediately that the stony brain was far larger than any baboon's. The skull's face lay buried in a separate chunk of limestone matrix. When fully revealed and the parts put back together, the head Dart had was that of a young child, one whose face was neither pure ape nor wholly human.

Though it took him decades to be believed, Dart knew he had the first concrete evidence that Darwin was right—that the human evolutionary birthplace was indeed Africa. Darwin based his prediction on the distribution of our closest relatives: Mountain and lowland gorillas, chimpanzees, and pygmy chimps (bonobos) are all exclusively African. Of the true great apes, orangutans alone live outside Africa.

Thanks mostly to the efforts of the colorful physician-turned-paleontologist Robert Broom, more specimens of Dart's *Australopithecus africanus* (African Southern Ape) emerged in the 1930s and 1940s. All came from the limeworks—deposits of pure lime and mixed debris (breccia) that had gradually filled up caverns dissolved into the ancient magnesium-rich limestone bedrock.

Dart thought his southern ape-men were endowed with material culture. He pointed to broken bones of various animals lying about in the cave deposits, claiming they were the weapons used to kill and butcher prey. Indeed, it was Dart who started the "killer ape" fantasy, giving Ardrey all the ammunition he needed for his artistic flights of imagination.

The truth tends to be more prosaic—perhaps the reason that stories of UFOs and the paranormal will always have a receptive audience. Paleontologist Bob Brain spent years carefully excavating Swartkrans, another of the Transvaal lime deposits. Far from corroborating the Dart/Ardrey version, Brain found that *Australopithecus africanus* was in reality the hunted, and not the hunter. Brain even found skulls with holes on either side—holes which exactly matched the position of the canines of a leopard's jaw. Leopards are fond of lying up in trees, where they stash their kills and consume them at their leisure. In the rather open countryside of the Transvaal today, isolated trees can still be found at the mouths of sinkholes leading down to limestone caverns below. The jumble of bones and other debris that make up the beds of breccia in large part seem to have fallen in from above—much of it undoubtedly the remains of leopard kills. It seems pretty clear that our ancestors were part of the diets of leopards, and perhaps other carnivores as well.

Who were these *Australopithecus africanus?* Standing around four feet tall, weighing in at something like one hundred pounds, these little creatures would probably have struck us as more ape than human in overall form. Nor would their brains have bowled us over. Anthropologists get a rough idea of brain volume from measuring the room inside the skull. And brain volume alone is only the roughest estimator of relative intelligence. These caveats aside, *africanus* had a cranial capacity of about 450 cubic centimeters (cc). Ours, in contrast, stands somewhere in the 1,350–1,400 cc range. Chimps measure in at just under 400 cc. Clearly, in terms of the all-important brain, with *A. africanus* we have a creature far more apelike than fully human.

Why, then, do we claim *A. africanus* among our direct ancestors? Primarily because they were up on their hind legs, walking. Their knees and pelvises in particular look much like ours, and not at all like a chimp's. Nor was *A. africanus* the first of our lineage to abandon four-leggedness. Upright, bipedal walking was the first anatomical modification, the key *adaptation*—the crucial and necessary first step to the eventual emergence of the distinctly and distinctively human condition. It was taken by the progenitor of *A. africanus*—a species known as *Australopithecus afarensis* (the Afar is a region in Ethiopia where many of these specimens have been found).

Donald Johanson's famous partial skeleton, known to one and all as "Lucy," is but the most famous of an entire series of East African specimens. Hands and feet show vestiges of the kind of curvature seen in tree-climbing apes—testimony that this primitive species had not yet altogether abandoned tree climbing in its normal, daily behavioral repertoire. Yet their feet, knees, and pelvises also reveal them to have been upright and two-legged. But the most stunning evidence of bipedalism came from Mary Leakey's discovery of the footprints of at least two upright hominids walking across a fresh, dampened deposit of volcanic ash at Laetoli, in Tanzania. The tracks are indisputably those of bipedal, upright hominids—distinctly more human and very different from any track left, for example, by a stumbling chimp forced to stagger a few yards on its hind legs.

Demosthenes was right: We *are* a featherless biped, and he hit the nail on the head on a most fundamental defining aspect of the human body. Long ago, anthropologist Sherwood Washburn suggested that bipedalism offers a means of looking out over the savannah for enemies. More crucially, it frees the hands to manipulate tools—including, at least eventually, weapons—and to carry food. More recent suggestions are perhaps even more on the mark: Upright posture, together with reduction of thickness of hair coating the body *and* development of additional sweat glands, would be an ideal means of combating the tropical sun on the African savannah.

The scenario is crucial to our story, for here is a case of an evolutionary adaptation, central in the annals of human evolution, that appears to be a direct response to major events in global climate—events that affected Africa very deeply indeed. The Laetoli footprints are 3.5 million years old. The Ethiopian and Laetoli remains of *Australopithecus afarensis* span the period from around 3 to 4 million years ago. Fossils between 4 and 5 million years ago, so far at least, remain frustratingly scrappy.

But there are a few things we do know. Five million years ago there was a sharp, marked drop in global temperature. Ocean levels shrank, suggesting the growth of polar ice caps—though so far there is no direct evidence of actual continental glaciation. Five million years ago, the Mediterranean dried up completely, when the Straits of Gibraltar temporarily closed through shifting motions of continental plates. The Mediterranean became a dry salt pan more than a mile deep. Savannahs spread on neighboring continents, in Africa at the expense of the forests that had covered nearly the entire continent.

No one should be surprised that evolution mirrors climatic change. That's the way it is supposed to work. Paleontologist Elisabeth Vrba reports that the seven major groups of antelopes that still dominate the African plains arose abruptly 5 million years ago, following directly on the heels of a spasm of extinction that took many other species away forever. That hominids apparently gained bipedalism as those savannahs spread seems plausible—an evolutionary adaptation as forests shrank and a species of ape ventured out to live life on the open plains. The forests were left behind to the ancestors of the modern African forest dwellers: chimps, bonobos, and gorillas.

But mark well that cooling pulse, the radical change of ecosystems, and the (presumed) origin of hominid bipedalism just under 5 million years ago. There were a few subsequent, similar episodes, when global climate change affected human evolution, driving some species extinct and prompting the evolution of others. The early stages of human evolution show our early ancestors evolving and becoming extinct in virtual lockstep with other, profoundly nonhuman species: pigs, elephants, antelopes, and virtually all other denizens of the changing African ecosystems. Only when culture began to dominate the hominid approach to the environment was the pattern broken—and the extinction of old and evolutionary emergence of new hominid species, including our own, was no longer a simple, direct response to episodes of climatic change.

The Ethiopian *Australopithecus afarensis* seem to have been living near open, standing water—possibly a large lake. There would have been plenty of trees in that environment (indeed, the very trees, their feet seem to be telling us, that they still resorted to from time to time).

One remarkable discovery of the so-called first family consists of the partial remains of some thirteen individuals—perhaps all taken by surprise and drowned in a flash flood. Though by no means definitive, such a fossilized gathering is at least consistent with the idea that these early species of hominids were living in small bands.

At the risk of trying to have it both ways—after taking a shot at playwright Robert Ardrey's nonprofessional assessment of the killing instincts of *Australopithecus africanus*—I must say that the scenario of ancient hominid life depicted in the Stanley Kubrick/Arthur Clarke film *2001* was not, in its barest essentials, all that far off the mark. In their movie, Kubrick and Clarke had small bands of australopithecines roaming the countryside, huddling for protection together at night— and occasionally meeting up with rival bands, with whom they fought for drinking rights at the water hole.

More, Kubrick and Clarke depict the invention of weaponry: A member of the defeated band, idly banging bits of bone around, discovers the powerful effect a femur of some large mammal has when he directs the blow squarely down on some other bones. This invention is transmitted to band mates, and soon they become triumphant at the water hole, killing one of the leaders of their rivals.

Fanciful, but within the bounds of reason. Bones fashioned as tools and weapons hold up far less well than rocks. Dart's putative bone tools turned out to be fragmentary leavings of the meals of porcupines and hyenas. On the other hand, chimpanzees occasionally use wooden tools, teasing termites from their nests and teaching others how to do the trick. We know of no definitive evidence that these early australopithecine species had material culture. But it would be surprising if they possessed absolutely no form of material cultural implementation whatsoever.

What did they eat? If the early australopithecines were not the hunters that Dart and Ardrey made them out to be, they were probably not obligate vegetarians, either. The teeth of *Australopithecus africanus* have the characteristic patterns of wear and surface scratchings that are the telltale signs of fruit and leaf eating. But the general pattern of the teeth is not all that much different from our own, and we are omnivores. Chimps will devour the occasional small mammal crossing their path that they are fast enough to grab. There is every reason to think these early australopithecines did much the same. They may have scavenged as well; some anthropologists have even suggested that they might have climbed the trees and raided leopard kills while the beast was out roaming.

Australopithecus afarensis, and perhaps especially *Australopithecus africanus*, have all the earmarks of the ecological generalist about them. Apart from their upright posture—a true evolutionary innovation and a true ecological specialization—and apart from their slightly expanded brains, there is nothing remarkable about their very generalized hominid anatomy. *A. africanus* is a "gracile" species, meaning lightly built, without the massive bones of the earlier presumptive male *A. afarensis* or the massive jaws and skulls of some later species of hominids. Massive skulls, jaws, and teeth—beginning with gorillas and including the "robust" australopithecines that show up *after* these early ones lived—are almost always associated with a purely vegetarian diet: bamboo shoots for gorillas, and nuts, tubers, and other tough veggies for the "robust" australopithecines. Negative evidence, in a sense, but I take the very gracility of the *A. africanus* skeleton as mute testimony to a highly generalized and omnivorous ecological lifestyle.

Small bands, gatherers, scavengers, perhaps hunters—it sounds familiar. There are still groups of our own species, *Homo sapiens*, who, despite tremendous leaps in cultural capacity in human history over the past 2.5 million years, retain an essentially similar approach to nature, specifically to the local ecosystem—as did these early australopithecines 3 million years ago. Each australopith population was integrated into its local ecosystem—relying on some species of animals and plants for food (and perhaps other items), competing with some for food and perhaps shelter, fearing other species as predators or simply as health hazards, while maintaining a neutral stance to others.

And there is no better proof of the contention that our ancestors 3 million years ago were very much integrated into local ecosystems than the events starting some 2.7 million years ago, when once again sudden climate change pulled the rug out from under African ecosystems. As a result, species disappeared and new ones took their place as ecosystems were rebuilt along somewhat transformed lines. This time the data are better, and we can see what happened to our own ancestors in a period of great climatic flux.

TRANSFORMATIONS

Something momentous in human history happened about 2.5 million years ago. Our ancestors started using tools—fashioned of stone, a material that leaves an enduring record for archeologists to spot. Chimps fashion disposable wooden tools on an ad hoc basis, and the early australopithecines may well have had the casual use of wood and bone. But prior to 2.5 million years ago, there is no convincing evidence of the systematic fabrication of material culture.

These earliest stone tools were modest artifacts. It took some time before archeologists became convinced that the odd assortment of stones first noticed by the Leakeys at Olduvai Gorge in Tanzania were in fact true tools. They look like stream-worn cobbles with their edges knocked off—the sort of broken rock one might stumble upon virtually anywhere.

But these "Oldowan" pebble tools are not the random workings of inanimate Nature. They were fashioned by protohumans in a primitive but systematic fashion. Someone would take a chunk of lava stone and bash it with another rock, spalling small flakes off the central "core." Most of these early tools were fashioned from basalt—a dark fine-grained igneous rock common in the volcano-rich East African rift system which began to form some 5 million years ago. Archeologists first suspected that it was the cores themselves that were the object of the work: more sophisticated tools—hand axes formed by striking flakes of material off central cores of stone—have been found since the nineteenth century.

Mary Leakey and other archeologists have documented a rich assortment of tools of this early Oldowan cultural type, and it seems pretty clear now that the flakes knocked off these cores were at least as useful as the cores themselves. Microscopy has revealed scratch marks on these flakes, marks consistent with their use as cutting tools—some for grasses, others perhaps for hides and meat. The central cores no doubt were also used for less delicate chopping tasks.

How simple, yet how momentous! Here is the first concrete evidence that ancestral humans were using their brains to approach the basic tasks of survival, of eking out an existence, and avoiding falling victim to other species out to make *their* livings. Unlike the Kubrick/Clarke scenario of the original "invention" of tool use in an animal boneyard, here we have definitive evidence of cultural *tradition*: These unassuming Oldowan pebble tools bespeak not only the cleverness of original invention but the very essence of human culture—the transmission of knowledge gained.

Who made these tools? Louis Leakey had been scouring the promising, but for the longest time stubbornly unrewarding, freshwater deposits exposed at Olduvai Gorge since the 1930s. Raymond Dart hadn't yet fully convinced the scientific world of the importance of his South African australopithecine specimens when Leakey got his operation under way in present-day Tanzania. The tools had been found, along with literally tons of animal fossils, but no humanlike bones until Mary Leakey found the famous skull her husband Louis promptly dubbed *Zinjanthropus boisei.*

Could Zinj have made the tools? Dated at about 1.8 million years, the Zinj skull was quite a bit younger than the earliest stone tools, which stretched back to about 2.5 million years. Then, too, there was the disconcerting fact that Zinj looks more primitive, more apelike than the older gracile australopithecine fossils. True, the braincase of Zinj came in at about 530 cc—bigger than *Australopithecus africanus*; but the skull is massive, with a huge sagittal crest very like that of male gorillas. The teeth—especially the molars, the grinding teeth of the cheek region—on the skull are truly enormous. The sagittal crest supports a huge muscle mass, needed to crush and grind the tough, gritty vegetable foodstuffs that were evidently the obligate element of Zinj diet.

Zinj doesn't fit into the preconceived notion of progressive "hominidization," something that would help bridge the gap between the gracile *Australopithecus africanus* on the one hand and ourselves on the other. Archeologists and paleontologists in South Africa, meanwhile, had already unearthed a series of fossils that seemed, while less massive, in the main similar to the Leakeys' Zinj. These were called *Paranthropus robustus*—"robust near-man."

We now know there were a whole series of species of these "robust" australopithecines. Together, they make a great object lesson in ecological specialization and its effects on evolutionary rates. The entire *Paranthropus* lineage lasted a total of only one million years, yet there were at least four different paranthropine species living in eastern and southern Africa during that time. Evolution works in large measure by branching, by the division of one species into two (occasionally more) descendant species. The rate of origin of new species, hence the rate of evolution, among paranthropines seems high compared with earlier and later rates of human evolution. The reason seems to lie squarely in their basic ecological adaptations: As obligate vegetarians, paranthropines were the quintessential ecological specialists—far more than their ecologically (and anatomically) more generalized australopithecine forebears.

Ecological specialists seem to speciate more rapidly than their generalist relatives. The reasons are complicated and not fully understood. I think that newly evolved ecological specialist species have a better chance of survival than generalists for the simple reason that they can focus on a narrower range of foodstuffs, thereby running less of a chance of losing out in competition with the superior numbers of its parental species. Fledgling species have a better shot at survival if they are doing things a bit differently from their ancestral species. Within any one evolutionary lineage, there are always many more specialist species than generalists—as Elisabeth Vrba has pointed out, for example, in African antelopes. Impalas are a single, generalist species, related to the six or seven living species of the highly specialized wildebeest-hartebeest group. Over the last 5 million years, there have been a total of only two or three species of impala, while we know of at least thirty-five to forty hartebeest-wildebeest species over the same time span.

The flip side of the ecological control of evolutionary rates is that specialists are also far more prone to extinction than generalists. The robust paranthropines bear this out nicely, too: None of the species appears to have lasted very long. The earlier australopithecine species—and at least one later species, *Homo erectus*—each had evolutionary histories of much longer duration than individual robust paranthropine species. And the entire paranthropine group was gone within one million years.

All of this has a direct bearing on our evolutionary future. The good news is that generalist species are relatively extinction-resistant. We are ecological generalists. We have achieved this state partly through direct heritage from gracile, generalized ancestors, and partly through the back door by acquiring an extreme specialization—a greatly expanded brain and the capacity for invention and learning. Culture, an extreme evolutionary specialization, ironically confers great ecological flexibility and generality on us. And this actually bodes well for our own chances in the future.

But we are also beginning to see, as we peruse the patterns of history of our remote ancestors, that evolutionary change is very much a function of the appearance of new species. Nothing much seems to happen, in an evolutionary sense, until a new fledgling species buds off from its ancestor. And *that* happens only rarely, most often during times of environmental stress—indeed usually only when ecosystems have been *so* stressed that many species already living in a region have been driven to extinction. Evolution and extinction, it seems, go hand in hand.

So who made those Oldowan stone tools? Bob Brain has unearthed evidence that the South African paranthropines may have indeed used fragments of leg bones of various species to help unearth the bulbs and tubers on which they evidently dined. Some bony shafts found in Swartkrans deposits are shiny, as if they had been inserted into the soil as the paranthropines dug. Recently, anthropologist Randall Susman has argued that paranthropines had the same sort of opposable thumb that we have and would have been quite capable of grasping tools. Nonetheless, there is no convincing association between paranthropines and those stony tools of Olduvai Gorge.

Paranthropines, it turns out, were not the only kind of protohuman living between 2.5 and 1.6 million years ago. And this is in itself exciting, for it links human evolution far more closely to the normal patterns of evolution seen in all other animal lineages. Because evolution resides in greatest measure in the multiplication of species, it really should come as no surprise that at one time or another during the past 5 million years of human evolutionary history, there were more than one protohuman species living at the same time. We, today, stand as a single, ecologically generalized, virtually worldwide-distributed single species. Back in the Upper Pliocene, between 2.5 and 1.6 million years ago, there were two distinctly different lineages of protohominids, both derived directly from the earlier gracile australopithecines. One lineage comprised the three or four species of paranthropines. The other, with one (possibly two) species, was the maker of those stone tools—and our ancestor.

Homo is the Roman word for "man," in the sense of "mankind," rather than male human (which was *vir* in Latin). *Homo* means "human." And there is, at least in this case, much in a name. Unable to attribute the stone tools—or, more critically, direct human ancestry— to their robust Zinj, Louis and Mary Leakey continued the search for the makers of those stone tools. Not long after the 1959 Zinj find, the Leakeys began producing hominid fossils in significant numbers— reward for the patient zeal through all those bleak years when nothing in the hominid line was showing up.

One particularly intriguing Olduvai Gorge find was the bones of a hand. To Louis Leakey, that hand looked distinctly human. These and other fossil remains were attributed to a new species, *Homo habilis*— "handy man." Leakey speculated that this was the hand of man that had fashioned the Oldowan stone tools.

The idea that there was a hominid species over 2 million years old that was enough like us to warrant the name *Homo* was not immediately popular. Louis Leakey was by no means an academic insider. His initial claims tended to provoke skepticism; his somewhat overbearing personality did not endear him to the profession at large; and his eventual successes, it seems fair to say, eventually evoked both envy and a somewhat grudging respect.

It was not until the Leakeys' son Richard began prospecting the extensive deposits along Lake Turkana (formerly Lake Rudolf) in northern Kenya that the very idea of *Homo habilis* gained the full measure of scientific respectability. In the 1970s, Richard discovered the famous skull known formally by its catalog number KNM-ER 1470 (i.e., Kenya National Museum–East Rudolf specimen number 1470). Piecing the skull together, the younger Leakey and his team calculated the brain size of their skull at approximately 750 cc—something of a leap forward. The elder Leakey's *Homo habilis* seemed now much more of a sure thing.

Enough material has since surfaced in South Africa, Olduvai Gorge, and the Lake Turkana region to assure us that essentially gracile hominids, with brains ranging in size from just over 500 to 750 cc, were living alongside the very different vegetarian robust paranthropines. Some of these "habilines" look very much like slightly bigger-brained *Australopithecus africanus*; others, like Richard Leakey's KNM-ER 1470, seem considerably more advanced. Some anthropologists are beginning to think that there were two different species of gracile hominid—*Homo habilis* and *Homo rudolfensis*—between 1.6 and 2.5 million years ago in eastern and southern Africa.

There is little doubt left in anyone's mind that these early habiline hominids were the makers of those crude yet so important stone tools. Skull fragments have now been identified as habiline dating back to the time of the very appearance of those stone tools: 2.5 million years.

Two and a half million years. That number keeps coming up. A robust paranthropine—the most primitive yet found and the apparent ancestor of all later southern and eastern African species—has recently been dated at around 2.5 million years. The tools and habilines show up then. And that's just about the time when the gracile *Australopithecus africanus* disappeared.

Something major happened around 2.5 million years ago, something that profoundly altered the course of human history and the very ecological fabric of Africa and much of the rest of the world.

CLIMATE, EXTINCTION, AND EVOLUTIONARY RESPONSE

What happened 2.5 million years ago? The story has been pieced together by paleontologist Elisabeth Vrba, who wondered why so many species of antelope, as well as so many other African mammals including human ancestors, drop out of the fossil record in a relatively brief interval between 2.7 and 2.5 million years ago. And why, just afterward, a panoply of other species show up. The southern and eastern African ecosystems changed profoundly 2.5 to 2.7 million years ago, and that change encompassed our early ancestors.

Vrba felt sure that a profound and rather abrupt climate change underlay the turnover of African fossil species starting around 2.7 million years ago. Soon after she made her prediction, paleoclimatologists began to see an abrupt worldwide drop in temperature starting about 2.7 million years ago—a drop that Vrba estimates to have been between ten and twenty degrees Fahrenheit. It turns out that the shells of calcium carbonate ($CaCO_3$)–secreting organisms take up different isotopes of oxygen preferentially: The ratio of O^{16} and O^{18} varies according to the ambient temperature when the organism was alive; relatively colder climes increase the amount of O^{18} relative to O^{16} present in the shell. This simple fact presents paleoclimatologists with a thermometer from which they can read ancient ocean temperatures. Most of the shells found in deep-sea ocean sediments belong to microscopic organisms that, in life, float near the surface of the oceans. Their shells record the surface temperature of oceanic waters and thus bear a close relation to air temperature. We can tell how cold or warm it was with amazing precision going back millions of years—and that's how we know of the great cold snap that began 2.7 million years ago.

That was the key: All those antelope and other species that Vrba saw abruptly dropping from the fossil record, and all those other species that seemed to replace them right away, were reacting to a global cold snap. And our ancestors and collateral kin were right there with them: The gracile australopithecine A. *africanus* is one of the contingent that drops out. And the habiline and paranthropine lineages show up just on the other, upper side of that line. Whatever was going on with the ecosystems generally was affecting our ancestors as thoroughly as it was every other element of the African biota at the end of the Pliocene.

Geologists have known since the middle of the last century that the earth's climate went through some wild swings in what has been called the Pleistocene Epoch—more familiarly known as the Ice Age. Calibrations of the last few million years of earth history have become quite precise of late, and we now know that the date of onset of the first of the four major glacial pulses of the Pleistocene came 1.6 million years ago. Radiometric dates are subject to some error—and revision. But this date of 1.6 million years, coinciding with the end of the Pliocene and beginning of the Pleistocene (the "Plio-Pleistocene boundary"), is unlikely to see much significant revision.

That means there was a cold snap 1 million years earlier that no one had ever detected before. Geologists have gone back and looked, but there is still no evidence that the polar ice caps grew, or that massive sheets of ice typical of the Pleistocene proper spread southward over huge areas of North America, Europe, and Asia—the way they were later to do four times, beginning 1.6 million years ago.

The effect was immediate on all the world's continents: The climate grew not only colder but distinctly more arid. Steppes and deserts appeared in the higher latitudes. And the tropical rain forests that still covered much of Africa even after the earlier cooling pulse around 5 million years ago rather suddenly gave way to the open grassland savannahs so typical of large regions of eastern Africa today. Palynologists—experts on fossil pollen—have documented the dramatic change in the plant cover of the region, as the wet woodlands retreated and gave way to drier grasslands dotted here and there with woodland copses.

With the plants—base of the food chain and prime determinants of the shape of the landscape in which animals dwell—so go the mammals. Mammalian species adapted to life in dense tropical rain forests are simply not comfortable out in the sun-drenched grasslands. We have already seen how true this was for members of our own proto-human lineage, back in the days around 5 million years ago when our ancestors split off from the other apes, who remained with the shrinking African forests. When the plants, backbones of any ecosystem, change suddenly and radically, it's curtains for the vast majority of whatever mammals might have been living there.

Vrba calls the African ecosystemic events between 2.7 and 2.5 million years ago an example of a "turnover pulse." Tension mounts as a status-quo ecosystem begins to absorb the shock of yearly decline in average temperature. Things might go along more or less all right for a while—but a threshold is reached sooner or later. Then all hell breaks loose. Plants that had been there a generation ago stopped reproducing and were never replaced. When they went, so did all the mammals who depended on them. That's the picture: ecosystem decline, as species ceased to flourish as vigorously as they once had; then a fairly quick kick-over as the old species disappeared and new ones just as quickly came in their stead.

What is the disappearance of the old and the sudden appearance of the new species all about? Is it sudden death, total annihilation of species, and the rapid evolution of new ones to take their place? For that's what the fossil record seems to suggest—at least when taken at face value and read only in one region.

In reality, the story is far more complicated. We now know that when the climate changes—when, for example, we see a sudden, dramatic drop in global temperature—ecosystems tend to be shuffled and moved about. Plants are the critical element in all of this, of course. And, at first thought, it might seem strange to think of trees picking up roots and marching off to some other region. But that's exactly what they do—not, of course, by stalking across the landscape but by sending out their propagules, their seeds. And if plants can do it, you better believe that all sorts of animals, certainly including mammals—and especially including humans—can do it.

On a larger scale, plants and the animals dependent on them shift ground as climates change. Each time the continental ice sheets crept south during the last 1.6 million years, the tundra collapsed south in front of the ice. So did the northern boreal forest and the more southerly mixed hardwood forests. When the ice retreated back northward, everything edged back north too, following the ice and responding to the global warming. Yet it is not all neat and tidy; ecosystems do not survive completely intact as they move around. But move around they do.

And when they do—when species move around, keeping up with

familiar habitat—they are survivors. Conditions may have become intolerable where once that species lived comfortably. But ecological change does not spell automatic death. Chances are that a species will hang on as long as there is an appropriate recognizable habitat *somewhere* and as long as the species has the time and basic biological equipment to *get* there. A major point of concern within the modern conservation movement is that environmental change may be happening so quickly that species will not be able to track familiar habitat as that habitat moves around the landscape. And there is now the added concern that vast tracks of agriculturally modified land, not to mention suburbs and huge cities, lie as obstructions in the paths of ecosystems moving about to track climatic change. Nonetheless, the old patterns still persist somewhat today: The great northerly migration of many animal and plant species during the twentieth century in North America represents a direct expansion of species' ranges as they simply track a progressively warmer climate.

That's what happened to a number of the East African species: Tropical rain forests persisted in many regions of (predominantly east-central) Africa, and presumably many species formerly living in East Africa prior to 2.5 million years ago managed to hang on elsewhere. By the same token, many of the species showing up in the newly expanded grassland habitats of eastern Africa just after 2.5 million years ago presumably were already alive and well and living in similar habitats elsewhere. It is not always easy to know this with certainty since deposits of the same age recording different ecosystems are not always readily available.

But we do know that many species must have become extinct with that change in climate, and that still others, among those that first show up in the fossil record just after 2.5 million years ago, most certainly evolved at that exact time. Many of the old species of antelope, pigs, rodents—mammals of just about every common African group—

simply died out. And new ones evolved, taking their place in a radically reconstituted eastern African ecosystem. And this most certainly includes the species of our own lineage, for the early gracile australopithecines disappear forever, and both the paranthropines and habilines appear just as abruptly.

Further discoveries may alter this picture, but that time interval of 2.7 to 2.5 million years pops up with such monotony, group after group, that surely we are seeing some real signal here. And so far the evidence is convincing that our own evolutionary history was very much controlled by events shaping change of the entire complex of African ecosystems 2.5 million years ago.

Extinction comes as habitats are disrupted and a species is unable to locate suitable habitat elsewhere. But the evolution of new species is also contingent on the fragmentation of old habitats. New species typically arise when populations of an ancestral species become isolated and evolve, often rapidly, to meet slightly new sets of environmental conditions. Our lineage had already been out there on savannahs, but apparently still living in the presence of significant numbers of trees. As the climate became still drier, and the grasslands spread at the expense of woodlands, the old species *Australopithecus africanus* relinquished its hold, but not without leaving descendants of at least two sorts, one lineage adapted to a wholly vegetarian life out on the plains and the other leading an ecologically more generalized existence, with an expanded brain and at least the rudiments of a material cultural tradition.

This was not to be the last time that an ancestral species of ours became extinct and a new one appeared in conjunction with a major climatic change. But it was the next-to-last time. That some of our ancestral species became extinct, and others evolved, in concert with a major environmental change shows us two simple yet profound things about the way we were: That our evolutionary patterns of extinction and evolution went in lockstep fashion with those of other species is prima facie evidence that—like all other species—our ancestors were integrated, as small local populations, into local ecosystems. It also shows us that we had taken only the most modest of steps toward "doing something about the weather"—to utilizing culture as our paramount approach to the physical act of living. All that was soon about to change.

THE ICE AGES: DOING SOMETHING ABOUT THE WEATHER

Ingredients for an Ice Age: Place a continent over one of the earth's rotational poles and wait for the earth's wobbly orbit to reach the point where energy streaming in from the sun is at its minimum. The earth's average surface temperature takes a sharp dip. Polar ice caps grow, and they are not melted back each summer. Antarctica's ice cap is still a major controlling element of our present-day climate. Cold waters emanating from the south polar regions sink deeply and flow north beyond the equator, profoundly affecting oceanic circulation patterns—and thus the world's weather.

The cooling pulses of 5 and 2.5 million years ago were harbingers of the famous Ice Age which began in earnest 1.6 million years ago. Once again, as temperatures dropped and the world's continental habitats became drier, dramatic events were unfolding in human evolution. If the earlier arrival of habilines was something of a great leap forward, the next step—just about at the time when the first of four mighty glacial waves began to engulf the northern hemisphere continents— was even more momentous: People not all that different from ourselves arrived on the scene. This was the last time that a species of the human lineage evolved in concert with a major environmental event, marking the clear beginnings of a shift to culturally based modes of coping with the environment.

Humanity, more or less as we know it, had appeared. There's no mistaking these early humans for apes. They stood over six feet tall, with limb proportions pretty much like our own. Kamoya Kimeu, field assistant to the Leakeys, made a startling find in 1984, at Nariokotome, west of Kenya's Lake Turkana. This time Kimeu really struck the jackpot: not just an isolated skull but most of a complete skeleton, soon dubbed the "Turkana boy." Anthropologist Alan Walker, who analyzed the remains, thinks the boy died at around the age of nine—in terms of modern human development, perhaps equivalent to a twelve-year-old. The boy stood at about five feet three inches tall and is estimated to have weighed just over 100 pounds. But Walker thinks that at maturity, the boy would have stood over six feet in height, weighing in at around 150 pounds—taller, if a bit skinnier, than today's averagely built modern human male (if, that is, there can be said to be an "average" physique among billions of anatomically diverse modern humans).

That boy died about 1.6 million years ago—about the time when the first big glacial pulse was transforming the landscapes of the northern hemisphere. An even older skull, from East Turkana, shows this early species of humans to have had a brain size of about 850 cc (once again, for comparison, ours measures in at around 1,400 cc). And there are several other finds that may be as old as 1.9 million years, suggesting that this advanced species may have evolved before the Ice Age was truly under way. If so—if the correlation between that first cold snap of the Pleistocene and the appearance of these advanced hominids proves to be not all that tight—then, of course, our lineage had already stopped evolving in discrete events in direct response to major climatic events. The data are just fuzzy enough, though, to be suggestive: It does look as if there is still cause and effect between climate, on the one hand, and the evolution of this new, advanced hominid species, this *Homo ergaster*.

Homo ergaster is hardly a household name. Much more familiar are the older names "*Pithecanthropus erectus*" and perhaps also "*Sinanthropus pekinensis*"—names, for example, that I had to learn in school in the 1960s. But at about that time, zoologists and paleontologists began peering over anthropologist's shoulders, suggesting in not overly polite terms that anthropologists should stop naming a new genus and species each time they unearthed an additional specimen. The point was well taken, and anthropologists fairly readily accepted zoologist Ernst Mayr's suggestion that all of the Mid-Pleistocene hominids—like *Pithecanthropus erectus* (the "erect ape-man" discovered at the end of the last century in Java) and *Sinanthropus pekinensis* ("Asian man from Beijing")—constitute a single species closely related to ourselves: *Homo sapiens*. Mayr's suggested name for this lot stuck: *Homo erectus*.

For years *Homo erectus* has been my favorite fossil species. That's because some of the best evidence of punctuated equilibria in all of the fossil record of human evolution is associated with this species. Punctuated equilibria maintains that once new species evolve, they typically do not show much subsequent evolutionary change. Species tend to be stable, if they are to survive at all—and *Homo erectus* lasted well over one million years, with little sign of evolutionary change during all that time. The point is crucial to understanding what lies ahead in the deep evolutionary future of our species—should we get that far.

When those African specimens (the Turkana boy, the famous skull ER 3733 that measures 850 cc, plus a few other earlier finds) first turned up, the natural tendency was to call them *Homo erectus*. But my colleague Ian Tattersall makes a convincing case that these earliest specimens should be regarded as a separate species—a close relative and probably ancestor of *H. erectus*, but nonetheless distinct.

At first I really didn't like the idea—admittedly because the longer the time span we can attribute to the existence of *Homo erectus*, the more striking is its apparent evolutionary stability. But these earlier "proto-*erectus*" specimens are critically different from later true *erectus*, and the reasoning (ironically) actually comes from a paper Tattersall and I published in the 1970s. Back then, we pointed out that true *Homo erectus*, with their massive bony skulls and faces, were hardly convincing ancestors for our own species, *Homo sapiens*. And it was Tattersall who identified a particular anatomical singularity in the skulls of later *erectus* that is not found in the skulls of any other hominid species, most certainly including our own. *Homo erectus* was universally considered the ancestor of *Homo sapiens* by virtual default, for it is the only known hominid species of the Middle Pleistocene. It now seems certain that our species arose from African ancestors—and *not* from the later, Asian *erectus* populations.

These beginning-of-the-Ice-Age specimens have the sort of gracile, light build that is reminiscent of *Australopithecus africanus*, *Homo habilis*, *Homo rudolfensis*—and ourselves, *Homo sapiens*. They make perfect ancestors for us, though there are but a few specimens so far found in the African fossil record between the last recorded remains of *Homo ergaster* (just after the start of the Ice Age), and the earliest fossils of undoubtedly modern *Homo sapiens*, which are about 100,000 years old. The way it looks right now, though, is that this intriguing species *Homo ergaster* (the name, incidentally, means "work man") is the ancestor to later *erectus*, as well as to the lineage that eventually led to ourselves.

But the greater height and bigger brains of *H. ergaster* are only part of the story. For these people had an advanced material culture as well: They were the inventors of the "Acheulian" industry—an assortment of hand axes and choppers which, for the first time, were executed with a recognizable, repeated style. These implements measured anywhere from six inches to a foot in length. The hand axes tapered to a point, whereas the cleavers had a broader chisel-edge to them. Both sorts of implements were fashioned from a central stone block, with chips removed from front and back to render what quickly became a traditional shape—a tradition that was to last for at least one million years.

That's the first essential point about these tools: They represent a stylistic tradition that is handed down from worker to worker, from one generation to another—not through genes but through demonstration and the user's cognitive capacity to learn. These stony tools harbor other fundamental lessons as well. For one thing, the Oldowan tradition did not disappear just because a newer, more sophisticated technology showed up. Sometimes new technologies do manage to drive out the old—simply because they represent a vastly superior means of getting the job done: Cars have just about eliminated the horse-and-buggy—though some cultures even in the United States cling to the older ways; guns drove the longbow to extinction pretty quickly in European warfare.

More often, newer technologies represent a lateral shift; television, for example, by no means drove out radio. The Oldowan tradition evidently persisted in some sectors to within a few hundred thousand years of modern times. Cultural innovation is not a simple matter of linear improvement; like biological evolution, cultural innovation happens stepwise, with new inventions cropping up in one region, perhaps later spreading elsewhere (guns were invented after Marco Polo brought knowledge of the requisite explosive powder back to Europe from China). And these new acquisitions often only partially displace the old ways. Cultural evolution acts like a ratchet, accumulating new approaches, but often not wholly eliminating older solutions to the same problem.

Tattersall points out another crucial aspect of the archeological record of cultural innovation: Invention of new technologies quickly gets out of phase with the biological evolution of new species. Recall that the habilines and the Oldowan stone culture appear to have arisen simultaneously. Stone tools go back to the very evolutionary beginnings of the genus *Homo*.

But we never see that pattern again. The real events in Africa attendant upon that initial glacial surge up north were the extinction of the other species of *Homo* (*habilis* and *rudolfensis*) *and* the abrupt appearance, at about 1.5 million years, of the Acheulian advanced hand ax/chopper cultural tradition. And that, at first glance, is remarkable. For we appear to be seeing the invention of a new technology, not coincident with the evolution of *H. ergaster* but several hundreds of thousands of years after *ergaster* first evolved.

But is this really so remarkable? Tattersall points out that you need a well-established system—an up-and-running species of *Homo*—before you can really expect to see much in the way of cultural innovation. And this is really exciting: for the pattern itself clearly shows that biological evolutionary "invention" and cultural innovation are "decoupled": The one does not depend on the other. They are in different domains—and the way is paved for culture to begin to dominate over biological evolution as a means of approaching the fundamental tasks of living—and of coping with environmental change.

There is a final joker in the deck. True, biological evolution is concentrated in the events leading to the emergence of one species from another—a sort of *between*-species phenomenon. Cultural evolution happens *within* species. Historically, the two seem to have had little to do with each other—after the initial appearance of stone tools and the genus *Homo*. But recall one fundamental similarity shared by cultural and biological evolutionary innovation nonetheless: Both depend on geography. New species arise through geographic isolation, and cultural innovation similarly arises locally—more often than not the work of single individuals living within small groups. Geography has played a huge role in both cultural and biological evolution, and this observation too has great implications for what might, and what might not, happen in the future.

Homo ergaster had something no other earlier hominid species ever attained: As far as we know, it was *Homo ergaster* that first gained controlled use of fire. Localized patches of burned clay suggest that *ergaster* had campfires. Fires suggest protection from nocturnal predators, warmth against nightly chills, and perhaps cooking. Cooking suggests meat, which in turns suggests hunting. Quite a few "suggests" here; there is nothing that definitively points to *ergaster* as an organized, systematic hunter. For example, there are no spear points in the Acheulian stone industry. But spears can be fashioned from sharpened bits of wood and bone—with little chance of showing up in the archeological record. It's frustrating, not knowing much about the dietary ecology of *ergaster*. We can only assume that *ergaster* peoples continued to gather plant products and perhaps to scavenge and even hunt. Certainly they continued to live in small bands. And certainly they were integrated members of good standing in the African savannah ecosystems of their time.

Controlled use of fire also suggests elements of communal social behavior that fit in well with the invention of stylistic toolmaking and the dissemination of this knowledge. Fires need tending, and perhaps embers were kept to start fires anew. Fires provide a focal point for many of a small band's activities, including especially the sharing of nascent cultural traditions. Was *ergaster* capable of a form of speech as we know it? No—not as we know it in ourselves: The frontal lobes of their brains were nowhere near as well developed as ours, and anatomists have concluded in any case that the human larynx, so crucial to the formation of the vast range of human utterances, was a relatively recent evolutionary innovation, confined in all likelihood to our own species. But *some* form of primitive verbal communication nonetheless must have underlain the cultural traditions and social behavior of *Homo ergaster.*

WILL THE REAL *HOMO ERECTUS* PLEASE STAND UP?

The start of the Ice Ages, then, had a major effect in both human biological and cultural evolution. We were still very much both victims and beneficiaries of climatic change. True *Homo ergaster* seems to have been well established prior to the actual onset of glaciation. But the habiline species became extinct just as the Ice Ages began. And *Homo ergaster* developed sophisticated tools and achieved dominion over fire just about the time *the* cold snap hit.

That there were as many as three species of *Homo* in eastern Africa between 2 and 1.6 million years ago is a bit mind-boggling. Ecologically, it is hard to fathom—especially since the gracile hominids, which include especially *H. rudolfensis* and *H. ergaster,* are generally interpreted as ecological generalists, the more so since the advent of complex material culture. As a rule of thumb, closely related species do not tend to occur in the same region—because their ecological requirements are so similar that one is sure to drive the other out. This is especially so of ecological generalists; specialists divvy up the ecological pie, focusing on slightly different aspects of the available resources. Generalists don't do that; they tend to work on a variety of different resources and do not, as a rule, accommodate competing generalist species in their midst.

However confusing the ecological picture presented by these early species of *Homo* might be, the paranthropines—specifically *Paranthropus boisei* in East Africa—did manage to sail right through the onset of the Pleistocene 1.6 million years ago. That specialized vegetarian species made it nearly to the onset of the *second* glacial pulse—some 0.9 million years back in time.

Homo ergaster, the early, gracile version of the classic *Homo erectus*, disappeared sometime between 1.2 and 1.4 million years ago. It was replaced in a direct evolutionary and ecological sense by its close relative and presumed direct descendant, true *Homo erectus*. Once again it is Olduvai Gorge that supplies us with the crucial information. "Olduvai hominid 9," a skullcap and partial face, has the robust build of later *Homo erectus* from Asia. The evidence seems clear: *Homo erectus* evolved in Africa in the early Pleistocene. It was the first species of our lineage to venture forth out of Africa.

Homo erectus had fire and the relatively advanced tool kit invented by its ancestor *ergaster*. When the second glacial pulse came about 0.9 million years ago, two things happened in evident response. In the Eurasian frozen steppes, many new species of large mammals appeared, evolving in response to that second cold snap. These are the classic Ice Age mammals of museum displays, children's books, and movies: the woolly mammoths, the mastodons, the woolly rhinos, saber-toothed cats, and giant bison. Many of these species persisted right through to the end of the Pleistocene. Their bones still litter the frozen tundra, and they are especially well-known from the famous La Brea tarpits in what is now downtown Los Angeles.

The second thing that happened 0.9 million years ago is that *Homo* *erectus* showed up in Europe and Asia.* That's right. Not only was *Homo erectus* the first of our kind to leave Africa, it did so in the very teeth of a major cold snap and glacial pulse. Fire was no doubt a crucial, enabling factor for this amazing move. Once again, there is no direct hard physical evidence that *erectus* was hunting those newly evolved large herbivorous mammals. But the coincidence is certainly striking. And life on the periglacial scene could not have been all that cushy: The arid savannah of Pleistocene Africa was surely a more luxuriant landscape than Europe and Asia in which to gather vegetable foodstuffs— and perhaps even to scavenge meats and capture small game. There's no way to prove it, but the large mammals of the second glacial pulse were probably a meaty attraction luring *erectus* out of its ancestral African haunts.

Whatever the motive, the movement out of Africa 0.9 million years ago marked a turning point in our relations with the natural world. To be sure, we still lived in small bands within local ecosystems—however briefly migratory bands might have stayed in any one region. But the time had long since passed when a new hominid species would evolve in response to a burst of climatic change, and to the extinction of its forerunners. Indeed, the time had even passed when hominid species routinely became extinct along with many other species whenever a

* At this writing, geologists Carl Swisher and Garniss Curtis, of the Institute of Human Origins in Berkeley, have just announced new and surprisingly old radiometric dates for two Javanese early hominid fossils: 1.7 and 1.8 million years. It may be that *Homo ergaster* actually left Africa around the start of the *first* glacial advance, rather than the second. However, the preponderance of archeological evidence—stone tools—in Eurasia still points to a correlation between the arrival of significant numbers of *Homo erectus* at the start of the *second* glacial advance some 0.9 million years ago.

major climatic episode occurred. *Homo ergaster* and *Paranthropus boisei* had even made it through the beginnings of the Ice Age. But this was something different: With the advent of the second glacial pulse, *Homo erectus* actually went *north* into the very teeth of the climatic storm.

We have already seen that some species respond to marked climatic and habitat change by simply locating suitable, recognizable habitat elsewhere. *Homo erectus* did this one better: It actively expanded the roster of possible habitats that humans had heretofore defined as "suitable."

There can be no question that it was culture, not biological adaptation, that allowed this move northward. The robust build of *erectus* has sometimes been taken as prima facie evidence of adaptation to cold climates. But Olduvai hominid 9 shows that the robust *erectus* build evolved in Africa, *between* major glacial advances. Nor did *erectus* oscillate between more and less robust states as times of glacial advance alternated with warmer climes.

It was culture, *not biological evolution*, that enabled humans to make that radical expansion in suitable habitat. It was the generality of our approach to nature, by then largely a function of our cultural kit bag, that enabled us to leave the familiar ecological setting of eastern Africa and invade wholly unfamiliar new habitat, with utterly unfamiliar species of plants and animals around us—and to be able to call it home.

The story of *erectus* throughout the bulk of the Pleistocene, down to at least 300,000 years ago, is a monotonous saga of success. Like nearly all successful species, *erectus* just kept on going, not changing much at all from its earliest progenitors in Africa right on down to the youngest specimens known: skulls from a cave near Beijing which prompted the original name *Sinanthropus pekinensis*. That all-important figure— cranial capacity—did indeed jump up once again with the evolutionary advent of *Homo erectus*. Olduvai hominid 9 measures in at around 1,067 cc. That figure, on average, was not to change for nearly one million years of subsequent *Homo erectus* history.

Nor did the Acheulian hand ax/chopper invented by *ergaster* change much in the hands of *erectus*. There *is* a bit of geographic variation to be seen, with the cruder implements characteristic of Asian populations and more highly honed implements coming from Africa and Europe. But that's about it. That is, after all, the mark of success. If it ain't broke, don't fix it. *Homo erectus* was a remarkably successful species. And it owed its expanded sense of suitable environment to the culture that it had largely inherited (as far as we can see in the archeological remains) from *Homo ergaster*. We had truly started to do something about the weather.

4

Becoming Human and
Stepping Out

We are still in the Ice Ages. True, temperatures have been on the rise
for at least a century now—though whether this is just another natural
climatic swing, or if indeed all the carbon dioxide and other green-
house gases generated by agricultural burning and industrial/home/
automobile hydrocarbon consumption is already having a global
warming effect, is difficult to say. But there is no way we can agree with
the early geologists who proclaimed an end of the Pleistocene and the
advent of the Holocene ("Truly Modern") period some 10,000 years
ago. All the ingredients for another glacial advance are still in place.

Waxings and wanings of continental glacial ice sheets in Europe and
Asia, and the correlated dry and wet spells in tropical Africa, provided
the crucial mix of changing habitats and periods of isolation that are
the basic grist for the evolutionary mill. Human species may no longer
have been falling to extinction or evolving in simple lockstep with
global temperature change. But the emergence of several human species, including our own, during the past 500,000 years nonetheless

reflects a profound interplay between environment, geography, and the basic adaptations of advanced hominids of the Upper Pleistocene.

The last 500,000 years of human prehistory have been very difficult to interpret, and there is still a lot of controversy of the most elemental kind plaguing the subject. Who, for example, were the neanderthals? Just a variant version of ourselves, or a separate, side-branch of human evolution? Even more basically, did modern "races" of humans evolve in place directly from differentiated *Homo erectus* populations, as some paleoanthropologists still claim? Or did *Homo sapiens*, our species, evolve like any other species since the beginning of time, by splitting off from an ancestral species in a single, particular region?

Theory, evidence, and common sense all support the latter: *Homo sapiens*, it now seems certain, evolved in Africa, perhaps around 150,000 years ago. And that means that human species evolved in and eventually migrated out of Africa at least *twice* in the course of history: first, when *Homo erectus* left Africa about one million years ago, quickly spreading as far east as Java; and again, perhaps just over 100,000 years ago, when anatomically modern humans began showing up in the Middle East, starting a trek that has long since taken us around the globe. And there is now reason to think there was yet a third foray out of Africa—after *erectus* but before we ourselves had appeared on the scene.

Homo erectus was a singularly successful species, one that remained essentially unchanged for nearly one million years. Like so many successful species before it, *erectus* had an effective ecological strategy, an approach mediated by technology, enabling *erectus* to survive and to do so without significant evolutionary tinkering on either its basic anatomy or its cultural kit bag. Brain size in *Homo erectus* hung on in the 1,000 to 1,200 cc range (averaging closer to 1,000 cc) throughout those million years.

But while *erectus* was hanging on up to about 300,000 years ago in

Asia, things were happening elsewhere. The fossil record in Europe and Africa is a bit spotty—and certainly a source of contention—in the period between 500,000 and 100,000 years ago. Most of the European and African fossils have brains a bit larger, and are somewhat more lightly built (gracile), than classic *Homo erectus*. It is for this reason that Ian Tattersall proposes they be recognized as a distinct species, *Homo heidelbergensis* (based on a jaw discovered long ago in Germany, the first specimen of this species to be found).

Brain size in *Homo heidelbergensis* was somewhat advanced over true *erectus*, averaging out at about 1,200 cc. Some individuals had higher values—up to 1,280, not too far below our own average of 1,350 cc. Perhaps *heidelbergensis* evolved from *Homo ergaster*, so far known only from Africa. In any case, *heidelbergensis* occurs in both Africa and Europe. The inference seems plain enough: *Erectus* had diverged from *ergaster* by 1.2 million years ago, and by about 0.9 million years ago had spread from its home base in Africa into Asia. At some later point, *heidelbergensis* evolved from African *ergaster*, and later also spread northward—this time into Europe.

Sometime during the course of its sojourn on earth, the species *Homo heidelbergensis* developed a more sophisticated stone toolmaking technology. Stones were selected and chipped to prepare a central "core." A last skillful blow would then dislodge a long flake, which could finally be finished into a true blade. Knives and spears—along with an impressive array of choppers, scrapers, and axes—had made their appearance. And, once again, the new technology shows up first in Africa, before it spreads to points north. Although the evidence is still scant, it seems likely that with *heidelbergensis*, humans had at last begun to become hunters of fairly large game.

What, then, of true *Homo sapiens*? Three lines of evidence make it quite clear that we indeed came from Africa, no doubt evolving from African populations of *Homo heidelbergensis*. A series of fossils from southern Africa, dating to about 120,000 years, provide the most convincing evidence—though a modern-looking skull from Ethiopia *may* be as old as 125,000 years. In any case, no "anatomically modern" human remains of such antiquity are known from anywhere else, and once again Africa appears to have been the birthplace of yet another hominid species: in this case us, *Homo sapiens.*

Molecular biology provides the second line of evidence of a common African origin for us all. DNA—that all-important genetic material containing the full instructions for developing and running an organism—is a "nucleic" acid. Most DNA is housed in the nucleus of each cell, but some DNA is also found in small organelles housed elsewhere in each cell. Mitochondria are the powerhouses of animal cells, deriving energy from nutrients to drive each cell's activities and thus all bodily functions. It is the presence of DNA in mitochondria (and in the chloroplasts in plant cells) that led biologist Lynn Margulis to the radical, but now widely accepted, idea that the complex cells of animals, plants (and even fungi) arose from the symbiotic fusion of at least two different forms of bacterial cells.

Each human child receives an equal complement of nuclear genes from its mother and father. Sperm cells are tiny, containing little but the nucleus; the mitochondria at the base of a sperm's tail power the long swim toward the egg. But the tail and associated mitochondria drop off as a sperm cell invades an egg. That means we inherit our mitochondria only from our mothers: The egg cell comes equipped with the full complement of all the basic cellular organelles—certainly including mitochondria.

Enter "Eve"—a great public relations coup as well as an important scientific conclusion developed by the late Alan Wilson and his colleagues at Berkeley. The Wilson group sampled an array of mitochondrial DNA from living humans around the world. They concluded that the diversity of mitochondrial DNA patterns converges on an African configuration—one that must have been present in an African *Homo sapiens* woman who lived sometime between 150,000 and 200,000 years ago. What better name than Eve?

Wilson's work has been sharply criticized. There were problems with samples and the statistical techniques used to compare the different genetic sequences. Yet in general the results seem to be holding up. And they agree so well with what the fossils themselves seem to be telling us.

The third line of evidence pointing to a distinctly African origin for *Homo sapiens* is a bit more circumstantial but nonetheless compelling. Many anthropologists agree that the persistence of the African megafauna (large animals, especially large mammals) implies that humans have long been in ecological equilibrium with Africa's ecosystems. In the fall of 1994, I was expounding on this very point in an ecotourism lecture in northern Botswana under a magnificent mangosteen tree, replete with its resident nine-foot black mamba, and in full view of herds of red lechwe antelope and elephant. Crocodiles and hippos lurked in nearby pools, and we had passed large numbers of African buffalo, zebra, impala, and wildebeests on our way to this Edenish spot. There were lion, hyena, and jackal nearby, and a leopard had been in camp the previous night.

This was in the Okavango Delta—a young, fault-bounded outpost of the great East African Rift Valley—home to so many early hominids and now the bearer of their bones. The Okavango looks very much like what Olduvai Gorge must have looked 3 million years ago—and for that matter like the Nile River Valley in Egypt until much more recently, to judge from sepulchral paintings of the Old Kingdom. As I began to speak of the spread of *Homo sapiens* around the globe—a story that links our arrival with local extinctions, especially of megafauna—it hit me that the persistence of the African megafauna bespeaks more than just long-term peaceful coexistence of humans and the African biota. It means we must have come from there originally—since every other place tells such a radically different tale: one of loss of species wherever and whenever we showed up. And that in itself is an important story.

There is little doubt that the earliest members of our species were still living in small bands, very much integrated into local ecosystems. Once again, the pattern repeats itself: There are no startling new cultural developments—at least of the sort that show up in the archeological record—that coincide with the arrival of full-blown *Homo sapiens* on the planet. But after a while, things began to change: Cultural innovations began to proliferate, and humans began to spread around the globe. And *this* time, as yet another human species set forth to colonize distant points, there really was a difference. This time, the world's ecosystems began to feel it; we began to have a direct impact on nature.

In an earlier book, *The Miner's Canary* (1991), I attempted to establish the simple reality of mass extinctions during the geological past—extinctions that, in one instance at least, may have killed off over 90 percent of the world's species. But that particular event happened 245 million years ago, 240 million years before anything even remotely resembling a true human being had yet appeared on earth.

I also argued that humans are at the root of the present-day "biodiversity crisis"—that our species, in other words, was beginning to destroy habitat at a rate and scale comparable to the mass extinction events of long ago. Whether it be from "natural" causes (without the helping hand of *Homo sapiens*), or whether it be through our own, however inadvertent, hand, extinctions are a direct function of habitat loss.

I argued that in our initial spread around the globe—invasions from Africa and the Middle East that began perhaps 50,000 years ago—we started, in a modest way, to disrupt ecosystems and to drive at least some species to extinction. That statement provoked some critical commentary: Some archeologists pointed out, for example, that it is now becoming somewhat fashionable to chastise the earliest settlers of the Americas—Native Americans—for their lack of ecological sensitivity. Some authors openly doubt what they take as the received story of a pristine situation, with many different peoples living in a stable relation with the natural world until the European conquest sent the natural order spinning out of control. In this relatively recent version of ecological history, pre-Columbian Native Americans are seen to have engaged in a number of practices that were not in any sense "good" for the local ecosystem.

We must be careful here. No group of people has ever engaged in overtly destructive behavior toward local ecosystems unless they had already begun to live *outside* those systems. Most of the examples given of harmful or destructive practices in pre-Columbian North America that I have seen, for example, pertain to Middle American agricultural land management procedures. And it is the advent of agriculture, to my mind, that provided the real break with traditional, within-ecosystem modes of life that a few groups of our own species still share with all other species on earth. Agriculture marked the beginnings of life outside those local systems.

What I had in mind in *The Miner's Canary* was something less dramatic. It seemed to me that as we spread around the world between roughly 50,000 and 10,000 years ago, still engaged exclusively in one variety or another of a hunting-gathering existence, we were coming on a bit too strong. Walking into new areas, it seems to me, modern humans became like bulls in a china shop. It is almost as if we had become too clever in honing our cultural adaptations to life as hunter-gatherers.

The evidence is fairly straightforward: Wherever we went, many other species seem to have become extinct shortly after our arrival.

Whether it was Malagasy peoples reaching Madagascar a scant 2,000 years ago, or peoples arriving on Caribbean Islands at about the same time; or people arriving in the New World 12,000 years ago; or aboriginal Australians getting to *their* home 40,000 to 50,000 years ago, the result always seems to have been the same: Substantial numbers of species soon disappeared, especially but not exclusively prime hunting targets, such as large game mammals and, in a few instances, large birds.

Thus it is significant that the *only* place modern humans did not drive a number of species to extinction was Africa itself: our birthplace, and a region in which humans were primordially in balance with the species they hunted. We didn't invade Africa; we were born there— and lack of substantial, human-caused extinction of species reflects that fact. It is only now that humans, through a combination of habitat destruction and overhunting, are pushing large species of African mammals to the brink of extinction.

The case is fairly plain that, beginning around 50,000 years ago, our ability to cope with the environment, to make an effective living as hunter-gatherers in a diverse array of circumstances, was reflected in our relation to the natural world. Even *Homo erectus* had had the cultural capacity to adjust to all manner of novel climates and ecosystems, with their unfamiliar foodstuffs, predators, and competitors. But there is no evidence whatsoever that *erectus* ever disrupted the systems that it invaded. But *sapiens* was—*is*—another matter entirely.

The first species that *Homo sapiens* drove to extinction was not a prey item at all, but rather a close ecological competitor. *Homo neanderthalensis*. For decades, anthropologists have considered neanderthals to be a "subspecies," sort of a pronounced "race," of *Homo sapiens*. Because they were massively built, with brains if anything larger than ours (neanderthal cranial capacity is a whopping 1,500 cc), the standard picture of neanderthals has been of an isolated group of

Homo sapiens especially well adapted to the climatic rigors of Ice Age Europe.

But that really won't wash. Neanderthals developed as a separate species from *Homo heidelbergensis*, right there in Europe, perhaps as long as 200,000 years ago. In contrast, we evolved somewhat later from African populations of *heidelbergensis*. And our different origins, naturally enough, show up in our different anatomies. Especially our brains: Though larger than ours, the neanderthal brain was differently organized. *Homo sapiens* has a more vertical forehead—mark of a greatly expanded cerebral cortex, the frontal lobes, literally the thinking part of the brain. Neanderthal brains are especially expanded to the rear—whatever that might imply; but their foreheads slope as did those of their ancestors, and it seems unlikely that they would ever have developed the cleverness that was suddenly to blossom within our own species beginning roughly 30,000 years ago.

There were other differences between these two species as well. Some anatomists and linguists have maintained that neanderthals, like their predecessors, did not have the fully modern form of larynx and so were incapable of the full range of speech of the modern Tower of Babel.

But I am especially struck by the observation of French archeologist François Bordes, who pronounced the rather exquisite "Mousterian" tools that were for the longest time the mainstay of the neanderthal stone tool kit bag as "stupidly" made. Bordes was not being rude; by "stupid" he simply meant that the neanderthals show no signs of individual creativity and originality. Neanderthal tools are the same wherever they are found—and *whenever* they were made, for the Mousterian prepared-core tools were made in the same way for at least 50,000 years. And, unlike *Homo sapiens*, neanderthals never imported from more than a few miles away the rocks from which they fashioned their tools. The tools were diverse, well made, and effective. But in the sense that they were virtually invariant, showing all the signs of simple slavish copying for millennia, they were indeed "stupid."

Neanderthal populations flourished especially during the last great glacial period, from about 100,000 down to about 33,000 years ago. Then something happened, and the neanderthals rather abruptly disappeared. That "something" was not a climatic event—unless by some stretch of the imagination the arrival of *Homo sapiens* can be construed as a climatic event. "Climactic" is more like it. Neanderthals and modern humans persisted side by side in Europe for a few thousand years, and there is even some pretty good evidence to suggest that neanderthals had begun to copy some of the advanced toolmaking techniques introduced into Europe by the upstart *Homo sapiens.*

But coexistence was not destined to last very long. There are no neanderthal skeletons preserved with *sapiens* spear points embedded in their rib cages (as there are paleo-Indian spear points in the ribs of a skeleton of an extinct bison species, found in 1927 near Folsom, New Mexico). Direct murder? No evidence exists, but we all know what has gone on throughout recorded times whenever one group of *Homo sapiens* invades the territory of another.

Perhaps simple competition was all that it took. As we have seen, two species, especially the sorts of ecological generalists that humans-conferred-with-culture so patently are, cannot persist for long in the same place. One eventually drives the other out, and the process need not take long. A few thousand years would do it.

Certainly there is nothing to support the myth that neanderthal blood courses through the veins of modern Europeans, or that a neanderthal, if clothed in current vogue, would pass unnoticed in a New York subway car. Neanderthals are gone, and our arrival was their undoing. We had become very good at invading ecosystems and exploiting them to the hilt, sometimes at the expense of competitors and major food items alike.

ANOTHER GREAT LEAP

Things were changing. True, we were still living in small numbers. All *Homo sapiens*, wherever they found themselves in an increasingly far-flung distribution around the world, were still living in fairly small bands. We were still living *inside* nature—meaning inside local ecosystems—however disruptive, even destructive, our initial impact might have been when we arrived and took advantage of newly found resources.

But things were changing nonetheless. The earliest evidence of a settled existence, complete with elaborate huts, hints at an early step toward detachment from the natural world. Fifteen thousand years ago, modern humans were constructing small villages in what is now the Ukraine. Their huts were built of scavenged mammoth bones—ingenious affairs in which stacks of immense lower jaw bones formed the sloping sides, while the doors were flanked with huge leg bones. Meat was kept in "lockers" in the permafrost, the permanently frozen ground, in what was then a tundra-like plain not far from the glacial front.

Food storage is a critical step in controlling economic fate. Squirrels, of course, store their food against times of want—meaning cold winters when nuts are hard to come by. Food storage in itself is hardly a declaration of independence from the natural world, but it is a vital aspect, and a necessary early forerunner, of full-blown agricultural practice. For starters, it means that you don't have to pull up stakes when the reindeer herd wanders away. Early signs of village life *before* the advent of agriculture show that settled existence was already an element of human life. Agriculture simply fostered an explosive growth of a lifestyle that had already begun to appear in isolated pockets as humans spread around the globe.

But there is something else about human existence between 30,000 and 10,000 years ago that is exciting, important, and at the same time tantalizingly hard to understand. It is the sudden appearance of—for want of a better term—human sensibilities. It is impossible to know when humans—our own species, or any of our closer relatives—acquired speech in the complete form with which we are familiar. It is still more difficult to judge when self-awareness or true consciousness appeared. But with the advent of stone toolmaking 2.5 million years ago, and with the invention of consistent, repeated *styles* of toolmaking, we have unmistakable evidence of both intent and a capacity to learn. How much was taught through verbal communication no one can tell. How much intent implies conscious self-awareness and an extended, conscious sense of purpose is also difficult to judge.

Still, beginning just over 30,000 years ago, there is indisputable evidence that humans were very much aware, *consciously* aware, of their environmental surroundings. For here we find the beginnings of art, commingled with an exploding array of consciously constructed artifacts designed to deal with an increasingly diverse set of economic tasks in daily life. Harpoons, fish hooks, needles, knives of all sorts—the economic armamentarium of Upper Paleolithic (Old Stone Age) life is light years advanced beyond the core-tool industry of even the earliest true humans prior to 30,000 years ago.

More arresting still is the adornment often found on these implements. In caves in still-frozen Europe, the discovery of full-blown art, some of it in spectacular color, removed whatever lingering doubts anthropologists had about the human status of these early people. What this art signified to the lives of its creators—what purpose it served—is still not entirely clear.

Images of large mammals dominate these European cave paintings, though human handprints and human forms (often in more cursory, stick-figure style than the eminently realistic bisons, horses, mammoths, and so forth) and even abstract symbols are also occasionally present. The stylistic form, especially of the large mammals, is invariably as sophisticated as any artwork produced in modern times. Indeed, some art historians believe they can recognize the distinctive work of individual masters—single individuals who may well have been responsible for most of the paintings in an entire region of caves—for example, at Altamira and other caves forty kilometers away in the Cantabrian Mountains of northern Spain. The existence of these artists implies division of labor within society, a harbinger of specialist sociocultural organization that became pronounced with the Agricultural Revolution.

But what is art *for*? Art for art's sake is the watchword in modern society, though aesthetic concerns are in fact complexly interwoven in all functional designs. Yet artists, whether visual artists or musicians, often resent critical evaluations of intent: about what, for example, a particular nonrepresentational painting is supposed to depict. Artists often maintain that their work is not necessarily about anything specific, let alone supposed to serve any particular function other than arouse a sense of pleasure or even of pain. Art, at least these days, is fundamentally about emotion; as an amateur musician, I have no trouble with this general line of thought about what art, any art, is "about."

But as an analyst of the past, I am also imbued with the conventional Western scientific tradition that holds that things happen for a reason, and that innovations—biological or cultural, in the case of humans—have some sort of functional significance. Archeologists have, quite naturally, thought of cave art—European, African, American, Australian—as having served some direct purpose in the lives of its creators. Not necessarily so, of course.

What function does the art perform? The old idea, long in vogue, is that it must have something to do with rites, and perhaps especially with success on the hunt. The paintings are often astonishingly deep in the recesses of these old caverns—far from light, in what must have been eerie, even downright terrifying conditions. Yet here came these intrepid souls, producing some of the most exquisite images ever wrought, rendered with the aid of a few sputtering animal-fat lamps. Few could have seen these paintings, and the poor lighting no doubt never showed them off to their full effect.

Yet there are good arguments against the sympathetic magic, good-luck-in-hunting hypothesis. For example, reindeer were the main hunted staple of the Upper Paleolithic larder, at least in western Europe. Yet reindeer are rarely depicted in the caves of Spain and France. Why expend so much effort on items that rarely, if ever, actually figured into the hunting plans of these early people?

Why do kids form such intense relations with dinosaurs? How come virtually every child knows the names of ten or twenty dinosaurs, when they lived, what they ate, and what killed them all off? The generic answer is simple: control. Kids are enthralled by those big monsters—which are conveniently and safely dead. Kids gain a form of control over the terrors of life—the imagined and the all-too-real—as they master the esoterica of the dinosaur world. Nor is this a new phenomenon: Dinosaurs "made" the American Museum of Natural History when they went on display in New York toward the end of the nineteenth century. Certainly dinosaurs loomed as large in the lives of my fourth-grade class back in the early 1950s.

But control over the powerful and threatening external elements of life is not the only sort of power kids typically derive from their wealth of dino lore. They also gain something of an upper hand over their parents at the same time. Kids can tell you the true scientific name of *Brontosaurus* (it's *Apatosaurus*)—an arcane piece of information their parents are unlikely to know. There is power in knowing things.

I am not suggesting that adult *Homo sapiens* of 20,000 years ago were like today's children. But I do think that children's use of dinosaurs is a vestige of a process that began long ago. There is pleasure to be taken in the ability to render clear, realistic images of objects. There is pleasure, as well, in simply viewing such renderings. There is also a sense of accomplishment, a pride in being able to create such images.

But there is something more: a sense of control over the natural world. Being able to talk about, to describe, to draw and paint a wild animal requires observation, thought, analysis, even intimate experience. It requires knowledge, and in human life, knowledge is power. It is the power of those with knowledge over those without, and it is power in the more abstract sense—by knowing something, we have in a very real sense some semblance of control over it. We are enabled: If that something that we know is threatening, we have taken the first step toward actually doing something about it. Medical research is the perfect modern example.

Cave art may well not have been about controlling the hunt. But it was nonetheless about knowledge and the power that flows from it: Power within the community—with the earliest known communal-level division of labor and inklings of an artisan "class." And power over the natural world in general. It was a first step—and a necessary one—to declaration of full-scale independence from the natural world.

Taking Life in Our Own Hands

There are a lot of parallels in life. Birds, bats, and flying reptiles sprang from three separate terrestrial vertebrate lineages. All three have modified forelimbs—"wings"—fashioned from the same basic set of vertebrate bones: a single upper arm bone (humerus), two lower arm bones (radius and ulna), plus wrist, hand, and finger bones. We, of course, as usual retain pretty nearly the ancestral vertebrate condition, including the original five vertebrate fingers—though our opposable thumbs, essential for grasping, are an innovation crucial to the human story. Birds, bats, and flying reptiles have highly modified forelimbs, each in slightly different ways, but all ending up with wings for powered flight—parallel developments.

If biological evolution regularly produces similar adaptations in only distantly related lineages, cultural evolution is positively rampant with parallelism. The same idea crops up time and again—very often at about the same time in isolated pockets around the world. Even now, though the world is linked in an incredibly intricate network of information exchange, independent efforts are continually churning out similar results—often driven by fierce competition to find new drugs, improved computer chips, or virtually any of the literally millions of technological items of the modern world. A slightly older example from evolutionary biology: the neglected work of monk Gregor Mendel was discovered no less than three times in 1900—triggering the birth of the modern science of genetics.

Agriculture is one of those ideas whose time came, and came, and came again. "Diffusionists" like to explain *all* cultural innovation as a bright idea that spread around the world. Mayan architecture, for example, was derived from the Egyptians—or so goes one particular diffusionist claim. For that matter, there are a few benighted souls who insist that the Egyptians got *their* knack for pyramid building from wanderers from outer space. Too often, the diffusionist view masquerades a disbelief that indigenous peoples in different regions could possibly have come up with something as sophisticated as a pyramid. Or agriculture.

Agriculture sprang up independently at about the same time in a number of well-identified, and undoubtedly isolated, places. The Middle East was first about 10,000 or 11,000 years ago, just slightly ahead of Africa, China, Southeast Asia, India, and the New World. In each region, raw material—different species of plants, and later, animals—were co-opted from local ecosystems. For that is what agriculture *is*: the removal of parts of a small subset of plant and animal species from the natural world, providing them with all their requirements as we manage their care and feeding. In the process of domestication, their features are transformed, becoming, often quite rapidly, ever more useful to their human managers.

As here and there we plucked local populations of a few plant and animal species from their natural surroundings, domesticating them, we of course transformed ourselves. We removed ourselves from the fundamental position in nature that we had heretofore shared with absolutely all other species since life began: we abruptly stepped out of the local ecosystem. We told Mother Nature we didn't need her anymore; that we could take care of ourselves.

It sounds a bit dramatic. But consider what it means to live *inside* nature—meaning to live in local populations within local ecosystems. It means dependence on the continued productivity of the system in which you are living. It means treating resources as if they are finite, instinctively and unconsciously (in the case of presumably *all* species save our own) moving on before food resources are depleted to the point of no recovery. Gorillas move on each day, avoiding depletion of bamboo stands and other foodstuffs in any one place. When elephants recently ravaged Kenya's Tsavo Park, an apparent exception to the rule, drought and, especially, the unnatural confinement of human-devised boundaries impelled them to consume every bush and tree in sight. There were simply too many elephants and too few available resources.

What does it mean to live *outside* ecosystems? It means that our interests no longer dovetail with those of the natural world around us (if ecosystems can be said to have "interests"). Inventing agriculture in a very real sense was tantamount to declaring war on local ecosystems.

Living *inside* a local ecosystem, some companion species are food resources, others competitors, still others predators or parasites. The vast majority have no direct effect on, no special relation, to you. They are just there, part of the background—though connections can be insidiously pervasive. For example, the one kind of tree that produces absolutely no edible nuts or fruits, whose leaves and bark yield no useful medicines, nonetheless anchors the soil, shades the understory, and perhaps intermittently shelters other species on which we *do* depend.

Agriculture changes all that. Even the "neutral" species in an ecosystem are now in the way. Weeds are simply unwanted plants. Before we domesticated plants, there was no such thing as a "weed." We still clear tremendous tracts of forest every year to create new arable and grazing lands. That is the war, the effects on the world around us that agriculture has wrought—and that has such threatening significance for our own future.

That is the significance of agriculture for the rest of the natural world—a significance that is beginning to double back and threaten us. But that's not the entire issue, nor even the relevant point at this juncture. For we have to consider how and why agriculture arose, and what its significance is for *human* life. We need to consider what it means that humans are living in this novel way—outside local ecosystems.

I am particularly drawn to a recent scenario involving one of my favorite peoples of the past—the Natufians who lived some 9,000 to 11,000 years ago in the region around the north end of the Dead Sea, in present-day Israel. Among the implements Natufians left behind were extremely small stone blades that archeologists figure to have been the teeth of sickles. The Natufians were harvesting grain. And they are the oldest culture so far found that may have been actually practicing a form of primitive agriculture.

Now, it is difficult to say with assurance that sickle-wielding people were harvesting grains they themselves had planted. Simple gathering of stands of wild cereals surely preceded even the modest beginnings of saving grains from one harvest and deliberately planting the next. But recent work, especially by archeologist Frank Hole and archeobotanist Joy McCorriston at Yale, suggests that the Natufians may well have had the knack of planting grains for the next harvest. And the scenario they devised fits in well with a major theme in human ecological history: the 2.5-million-year-old, culturally imbued search to take control, to do something about the weather.

We have increasingly minimized, contained, and controlled the direct effects of the natural, physical world on our lives. We have done much the same with the biological world. But, in the most important step in the process—the invention of agriculture—it should come as no surprise that, at least in the Middle East, that step was taken at the prodding of climate change! Or so think Hole and McCorriston, who have pointed to a period of hotter, drier summers that hit the eastern Mediterranean region around 12,000 years ago.

As the climate warmed, the vegetation we think of as typical of the Mediterranean—such as olives, cedars, and junipers—moved into the region, accompanied by a number of species of annual grasses. Traditional food sources—along with a number of lakes, where Natufians tended to locate their settlements—began to dry up. Key to the Holt-McCorriston scenario is einkorn wheat, one of the earliest cereal grains to be domesticated. Wild einkorn tends to lose its kernels easily, making harvesting a difficult, inefficient job. But some mutant varieties hold their kernels better, and the thinking here is that harvesting of einkorn would rapidly increase the frequency of the mutant form, because more and more of the mutants would have been harvested and kept over for next season's planting. Whether or not the details of this scenario stand up—and archaeologists have yet to find definitive evidence of domesticated einkorn in a Natufian settlement—the general sequence of events seems close to the money.

Domestication of animal species came just a bit later. Stephen Budiansky, a writer for *Newsweek* magazine who maintains a working farm, has concocted a novel—and absolutely false—theory about animal domestication that is nonetheless sufficiently ingenious that it actually helps shed a bit of light on this whole process of stepping outside local ecosystems. The conventional story, of course, is that humans manipulated certain tractable species of animals, effectively bringing them along in the process of domestication as we ourselves departed the confines of local ecosystems.

Budiansky has taken the conventional account and stood it on its ear. Animals, he tells us, domesticated *us*—effectively hitching a ride along with us as Pleistocene climates changed and so many of their fellow species began to fall to extinction as their old familiar habitats disappeared. Budiansky's not-so-hidden agenda is to counter criticisms of animal rights groups who object to farming practices.

But claiming that domestication is essentially a ruse, where sheep have pulled the wool over our eyes, is equally absurd. True, as Budiansky points out in his very best argument, Laplanders and reindeer both benefit from their close association, and it can be hard to say who is controlling whom as the Laplanders tag along after the migrating herds. Laplanders hunt the reindeer, who in turn tolerate their predators because of their fondness for Lap urine.

Budiansky was right to stress the dependence that each of the domesticated animal and plant species has on the human ecological course we first embarked on some 10,000 years ago. If it is true that we depend on them, then, we believe, we must have domesticated them. Budiansky thinks it is the other way around: It is they who depend on us, and therefore *they* chose *us* as a safe haven, the lesser of two evils as they otherwise ran high risks of extinction.

But however much reindeer have come to rely on humans, and however much they may actually seem to be taking advantage of the situation, it is of course overt conscious human cleverness that lies behind all instances of domestication: of plants, of animals, of ourselves. We picked up, in more or less the fashion suggested in the Natufian scenario, the ability to alter the behaviors of some of the creatures around us that were proving useful. This is absolutely crucial: Our ability to manipulate the genetically endowed properties of organisms goes back to the very beginnings of permanent settled existence. Genetic engineering may be a major wave of the future, but it has already been a core ingredient in the very founding of complex social systems and the rise of what (often for want of a better term) we continue to call "civilization."

By the same token, we must realize that, when we see that we ourselves have stepped out of local ecosystems, we are nonetheless not wholly unique; we have brought along a number of species with us— or so it seems at first glance. Many domesticated animals and plants are incapable of surviving even if returned to identical ecological surroundings from which they originally sprang. Indeed many are incapable of reproduction without direct human intervention. Nonetheless there is one very big difference between our relative independence, our change in status vis-à-vis local ecosystems, and the nature of existence of most, if not all, domesticated species.

The difference is this: Our manipulation of the genetics of plants and animals perforce focuses on small subsets of organisms—never entire species. We take local populations and manipulate their characteristics—originally purely through selective breeding, and now, increasingly, through direct manipulation of their DNA—their genetic information. The effects are prodigious, often impressive, and sometimes ludicrous. Many years ago, Russian geneticists crossbred cabbages and radishes, hoping to obtain a plant with the virtues of both; as fate would have it, they got something with the head of a radish and the root of a cabbage.

So the domestication of other species does not, indeed *cannot*, entail the modification of *entire* species. The original species, at least initially, still lives on as ever before in local ecosystems. When we control the properties of certain animals and plants, we are not controlling the evolutionary fates of entire species. We are plucking out parts of species, creating independent lineages, and molding, to a greater or lesser extent, the properties of these artificially culled lineages. We should bear this in mind as we contemplate the potential impact of genetic engineering on the human future. As the past has already so eloquently taught us, it is next to impossible to modify the genetic properties of an entire species—including *Homo sapiens*.

SEEKING DOMINION

We have reached the most critical juncture in human ecological history: the point where, 10,000 years ago, we became the first species to venture beyond the confines of the local ecosystem. Ecologically speaking, the past 10,000 years have been purely transitional, an epoch where the success of an agriculturally based approach to human existence went unchallenged. Despite drought-engendered episodes of famine (the apparent cause, for example, of the two major interregna separating the Old, Middle, and Late Kingdoms of Egyptian history), agriculture has been a virtually unmitigated success. The period is transitional because now we see the beginnings of serious, lasting threats—to the global system, to ourselves—if the same course is recklessly pursued for much longer.

Evolutionary ecologists tend to measure success by expansion of population numbers, and the explosion from a few million to 5.7 billion human beings in a scant 10,000 years certainly qualifies on that score. But the past 10,000 years, of course, are far more than a heady period of population expansion fueled by this novel approach to the basic necessities of living on Earth. It is the span that we think of as human history per se—all the glorious accomplishments, as well as all the wars and other social ills that have come along since we forsook local ecosystems for a rural agrarian, and increasingly urbanized, existence.

The symbolic legacy of this ecological revolution—the stories we began to tell ourselves about who we are, where we came from, and how we fit into the world—still grip our collective consciousness. Our modern, Western-world view was shaped to a very great extent in the earliest phases of the shift to agriculture and the origins of complex nation-states. We have to face the fact that old stories that were accurate and viable 10,000 years ago have outlived their usefulness, now that our no-longer-novel agriculturally based strategy is beginning to raise serious problems for our own future.

What people actually say about their world and how they fit into it is invariably revealing and usually, if obliquely, fairly accurate. Anthropologist Colin Turnbull reported that hunting parties of the Mbuti pygmies of the Ituri Forest of modern-day Zaire sing out to their surroundings, chanting greetings to "Mother Forest, Father Forest." I take these and similar ethnographic reports both literally and seriously. To me they reveal a conscious sense of actual situations. In this case, local populations of Mbuti are integrated (or were until very recently) into the Ituri Forest ecosystem.

That's why the earliest written texts, certainly including the Judeo-Christian Bible, are so exciting. We have a firsthand account of how the early agriculturists saw themselves in relation to the natural world.

Whatever one makes of the religious significance of the Judeo-Christian Bible, virtually everyone agrees that it yields a marvelous window into the thoughts and feelings of those who chronicled the times. In my debates with creationists, I have often said that the cosmology of Genesis represents the best thinking available at the time. It need not be true to be utterly fascinating. We now know that the earth is 4.5 billion years old—far older than the scant 6,000 years or so implicit in Genesis (as determined centuries ago by counting up the ages of the ancient ones, including Methuselah, in the "Begat" section of Genesis 5). It is no insult to the intelligence of the ancients that human knowledge has increased. Nor is our understanding today by any means a final lock on the truth; today's knowledge is every bit as much an interim progress report as were the pronouncements of Genesis.

Genesis tells us clearly what at least some people thought about how they fit into the natural world not long after agriculture had sparked the fundamental step outside local ecosystems. After creating heaven and earth and all living creatures, God said:

> . . . Let us make man in our image, after our likeness: and let them have dominion over the fish of the sea, and over the fowl of the air, and over the cattle, and over all the earth, and over every creeping thing that creepeth upon the earth.
>
> So God created man in his *own* image, in the image of God created he him; male and female created he them.
>
> And God blessed them, and God said unto them, Be fruitful, and multiply, and replenish the earth, and subdue it: and have dominion over the fish of the sea, and over the fowl of the air, and over every living thing that moveth upon the earth. (Genesis 1:26–28, King James version)

> And God saw everything he had made, and, behold, *it
> was* very good. (Genesis 1:31)

These words constitute the most ringing declaration of independence ever set down. They say that people, whatever their similarities with beasts of the field, are unlike any other living species. We are entitled to the earth and to all its fruits. We *own* the earth, and we must seek dominion over "every living thing that moveth on the earth."

I take this passage to be as simply and profoundly eloquent *and accurate* a statement of how the Israelites (and all others living in the newly agriculturalized mideast) fit into the natural world as the later Mbuti statement reflects a conscious and accurate assessment of how *they* fit into the Ituri system. According to Genesis, we are not of the natural world, but above it and lords over it. Yet if we don't fit into the natural realm, where *do* we fit? The answer lies in a fundamental change in the basic conception of the spiritual world.

The Mbuti's Mother and Father Forest are sometimes characterized as gods, as spirits of the natural world. Certainly the Mbuti of Turnbull's description are asking permission of some higher authority—of spirits, perhaps, but spirits as the abstraction of the forest, the ecosystem itself.

What do you do, then, when you step outside that system? You can no longer importune the spirits of the very system you have abandoned. Indeed, you are far more likely to deny their existence. They are the gods of a system that you now seek dominion over—over which you seek absolute control and in essence have, with tilling of the fields, declared war on. The ecosystem, that natural wild place that used to be home, has become the enemy.

What a heady feeling! Yet how scary! *We* became masters of our own fate, turning our backs because we were no longer dependent on the nourishing local ecosystem. It must have seemed as though we had co-opted the role of the gods. But that would hardly do. We had to take something comforting along with us—and that, I submit, was an abstract version of the ancient spirits of the wild. Though we claimed just the opposite, we invented gods in our own image, gods to render ultimate authority and the real control. Ultimately, some of us came up with the concept of a monotheistic God—still very much created in our image, and still very much the ultimate authority figure.

The Mbuti seem to be right about how they fit into their surroundings. Genesis reveals that the early agriculturists were equally correct about how they (didn't!) fit in. But neither set of descriptions fits *our* current situation at all well. We need to rethink how we do fit in as we careen off into the future. And this rethinking may well have theological implications, as it indeed did for the Mbuti and the early Israelites. As we become accustomed to seeing ourselves as still within nature— but as we realize that we are interacting as a single species with the entire global mega-ecosystem—we may one day revert to a concept of God more clearly tied to the earth itself. Some of us already have. But we should never forget that God helps those who help themselves.

5

The Way We Are

Culture, consciousness, cognition—humans truly are unique. We are sufficiently unusual as animals that the German biologist Bernhard Rensch once proposed that we be allocated to our own separate kingdom—the "Psychozoa"—on a par with the animal, plant, fungal, and microbial kingdoms.

Yet for all our preoccupations about how special we are, some of the more obvious uniquely human traits have gone largely unrecognized. The three I have in mind are our divorce from local ecosystems, our formidable inner-directedness, and our emerging status as the first species ever to interact with the global economic system as a whole. The whole earth has become our local ecosystem, and our entire species functions as a single, massive population within that system.

There is a pattern of historical contingency linking the three: We could not have become a single economic force within the global system had we not first stepped outside the confines of local ecosystems—and had we not had the culturally endowed means of communication that has enabled us to emerge as a single economic entity vis-à-vis the global ecosystem. We need to dwell just a bit longer on the factors that enable our internal cohesiveness, as well as the actual dynamics of life *outside* the local ecosystem—and *within* the planetary ecosystem. Thus armed, we can then contemplate our future.

REPRODUCTION, ECONOMICS, AND THE BODY POLITIC: THE ROLE OF SEX IN HUMAN LIFE

Yet another singular aspect of human life: the disjunction between sexual behavior and reproduction. Unlike all other species, humans engage in sexual behavior for more than purely reproductive reasons. Decoupled from pure function, sex plays a major role uniting economic and reproductive facets of human existence—and in so doing contributes enormously to our inner-directedness and collective self-absorption. Sex, strangely enough, has a lot to do with our economic lives, and thus is crucial to our position in the natural world.

Nearly all true animals—from corals up through clams, worms, fruit flies, and sea urchins—reproduce sexually. So do we, of course. But it is obvious to all that human sex goes far beyond simple baby making. Sex links our economic and reproductive lives in complex ways not seen in any other species. Sex lies at the very heart of human social behavior.

Nearly every organism leads two kind of lives: It gets energy to live (the "economic" side of simple existence), and, sometime in its lifetime, it at least tries to reproduce. This is true of absolutely everything—bacteria, trees, walruses, and people. When a mountain lion is out hunting, she is leading the economic side of her existence. She is concerned with jackrabbits hiding behind sagebrush (i.e., with *other* species in her ecosystem), but when she is in heat, she looks for a mate. The focus is then purely reproductive: She is out to make more mountain lions. Those two aspects of her life—eating and reproducing—are basically discrete. Economics and reproduction are largely independent realms in life.

But not so with social organisms. Sociobiologists love to pretend that the social behavior of honeybees is all about workers, who don't themselves reproduce, helping the queen and her drones make more honeybees. The workers help because they share a certain percentage of the queen's genes—and so, in a sense, get to reproduce vicariously. Social systems, to modern theorists, are fundamentally reproductive cooperatives.

But there is more to a honeybee hive than just baby-bee making. Most of the work those workers perform is purely economic: They are out foraging for nectar and coming back to tell each other where the goodies are. Those nectar supplies are the fuel that keeps each bee alive—and, to be sure, enables the queen to go about *her* purely reproductive business. For my money, honeybee hives are a fusion of the reproductive and economic interests of their constituent bees, with a few doing the reproducing and the rest the economic work—a division of labor that effectively fuses the economic and reproductive sides of every organism's life in interesting and complex ways.

The same is true of humans, but with some huge differences. Humans just don't divvy up tasks the way bees do in their hives. Each of us humans leads an economic *and* a reproductive life—though the option to reproduce is foregone in a variety of circumstances, and seemingly more often than in most other species. Though biologists like to insist that reproduction is the heart and soul of the very state of being alive, it is nonetheless true that *all* organisms *must* obtain energy to live. It is also the case that reproduction is the single bodily function that is *not* essential to keep an organism alive. That some organisms in some circumstances routinely decline the option to reproduce is astonishing only to those evolutionary biologists who insist that the drive to leave as many copies of one's genes to the next generation is the fundamental, overwhelming force and factor of life.

Humans aren't bees. We don't leave reproduction to a few members of a reproductive caste while the rest of us go about the task of earning a living. In that sense, our social existence is less of a departure from the more solitary lives of, say, mountain lions. Loners for the most part, individual mountain lions hunt and reproduce. So do we humans— though most of us now type on computers, mend shoes, or till fields rather than hunt for a living. The human social cooperative mingles economic existence with reproductive duties in entirely different ways than is the case with honeybees. And sex plays a big role in connecting our economic and reproductive lives.

Though they wrangled over the issue for years, anthropologists have generally conceded that nuclear families are ubiquitous in human cultures. Not to say that all families conform to the cereal-box stereo-type of the 1950s, with Mom and Dad, older sister and kid brother all happily wolfing down wheat flakes around the kitchen table at 7 A.M. There are other forms such families take—an issue of importance even in campaign rhetoric during recent U.S. presidential election years.

But there is definitely something to the union of male and female. After all, humans, like every other species, do reproduce. But such bonds, such nuclear families, are not formed for the exclusive purpose of producing children. They are also, and just as fundamentally, eco-nomic cooperatives. Just as with honeybees, human social systems at base (meaning the nuclear family base) are fusions of the reproductive and economic interests of the parties involved. Insisting that human families are purely reproductive in nature and "intent," designed solely for the purpose of having kids, is as distorted a view of human sociality as insisting that a honeybee hive exists only to make more honeybees.

But humans, of course, arrange things rather differently than bees do. Biology drives the division of the essential aspects of reproductive labor, with males doing little more than producing sperm at the opportune moment, while women nurture the developing embryo for the nine-month average. Despite aspects of changing male roles in early child rearing (at least in some segments of American society—a fairly narrow perspective), the overwhelmingly critical role of the mother in the first year after birth seems nearly as biologically imbued as the gestation process itself.

All humans, too, lead economic lives—the focus of much discontent in recent years as "traditional" roles, especially for women, are reexamined and redefined. Unlike the purely reproductive situation, there is no biologically driven mandate for economic division of labor among men and women. Even tasks conventionally equated with the (averagely) greater brawn and aggressive tendencies of males have been (however reluctantly) thrown open to women in increasing numbers in recent years in technologically advanced societies—as they have been, from time to time, throughout the millennia in one culture or another. (Egyptian pharaohs were supposed to be male, but Hatshepsut and Cleopatra are remembered as much for their successful managerial exploits as for the fact that they were, as women, anomalies in the Pharaonic system.)

Biology might not mandate it, but culture always does instill one form or another of male-female division of economic labor. It is our culture that is changing, embracing changing economic roles of women right now. Hunter-gatherers typically show at least partially different economic roles for men and women within nuclear families, as well as within the band generally. The Bushmen of the Kalahari send (*used* to send is perhaps more accurate) the men out on the hunt while the women search for edible tubers (while tending to the economic needs of the younger kids; feeding a child is a perfect example of a hybrid economic/reproductive activity). For their part, the men would train older boys in the skills of the hunt. On the other hand, Mbuti pygmy hunting parties involve(d) both men and women. Nothing is cast in stone, but at least because women typically retain charge of child care even beyond the years of infancy, it is mostly men who do the hunting, and women who do the gathering (or crop tending).

So there is something to anthropologist Helen Fisher's intriguing suggestion that nuclear families might have developed out of a food-for-sex deal. Human females are virtually alone among mammals in being fully sexually receptive 365 days of the year. Monthly menstrual cycles are not at all the same thing as the annual (or semiannual) estrous cycle typical of mammalian females generally, including all of our closest ape relatives save (intriguingly and suggestively) bonobos—"pygmy chimps." Estrous is all about reproduction: Non-human females come into "heat" at the time of ovulation. The whole idea is to become pregnant.

Human menses are something entirely different: the monthly discarding of the lining of the uterus, the very least likely time for fertilization and implantation to take place, as anyone who has used the rhythm method of contraception knows. Ovulation, thus maximum fertility, takes place in the middle of the cycle, two weeks or so out of phase with menstruation.

Sexual receptivity has nothing to do with menstruation, except for possibly actually depressing libido in some individuals some of the time. Why did female sexual activity become dissociated from ovulation? Sex began to serve a role in human life over and beyond simple reproduction. Fisher thinks that role was an exchange: Males could indulge in sex more than one or two times a year, and females could benefit from the economic ministrations of their now more-permanent partners. The idea hinges on the assumption that mammalian males primordially are willing to engage in sex whenever the stimulus is present—probably a safe enough notion. In suggesting that the primary female reward is economic, it also somewhat begs the question of *female* primordial interest in matters sexual.

But the idea does have something going for it, simply because humans take so very long to mature. Much has been made of the conflict between our very large-brained heads and the diameter of the birth canal. We simply could not go on forever expanding our brains in an evolutionary sense without running into serious problems with the birthing process itself. Expansion of the female pelvic girdle apparently is limited by the need to remain a fully functional, upright, and walking human being. The conflict was headed off at the pass by taking brain and head growth only so far in the internal gestation process. When a baby is born, the roofing bones of the skull are not yet knitted together. The brain continues to expand *after* birth, and the bones begin to knit much later.

In effect, part of the human developmental process goes on *after* a child is born. Nor is it just a matter of needing a period of helpless infancy while our brains continue to develop. Our increasing reliance on culture over purely physical attributes as the very basis for making our way in the world means that we have to spend a lot of time getting up to speed. Kids have to learn, and do so as they continue to grow— another good reason why evolution has slowed the rate of maturation, including reproductive maturity, to a much later age than is customary among our nearest great-ape relatives.

So the biology of big-brainedness and its corollary of learned culture require an increasingly long period of immaturity—one in which a great deal of care and feeding is required. A child's job, in a strict economic sense, is to know its parents well, the better to induce them to maintain this protracted period of care as the long process of physical maturation and cultural indoctrination goes on. That seems to be the main payoff of the very existence of nuclear families: The sex-for-food arrangement sets the stage for a stable pair bonding that in principle can last far longer than the ephemeral couplings of rutting mountain lions. One need not be a champion of "family values" to acknowledge the importance of such stability—including a regular paycheck—when it comes to rearing children. The basic economic-reproductive cooperative that is the stable pair-bond underpinning of the "nuclear family" depends on sex for its very existence. And that means sex liberated from pure reproduction.

If much of the work that goes on in a honeybee hive is indeed vicarious reproduction, with workers tending to the queen's larvae and bringing in nectar for all to dine on, human social structures are mostly about economics: work, or economics in the narrowly defined, human meaning of the term. No human social structures beyond the nuclear family are expressly designed for reproductive purposes, though (largely abandoned) arrangements at Israeli kibbutzim and similar experiments of social, collective child rearing come to mind as partial exceptions to this rule. Conventional schools are clearly economic entities that provide jobs to community members. They are, as well, a mechanism of enculturation that supplements parental care and thus aids the reproductive interests of the parents. Schools are a vital adjunct in complex societies where no single individual knows enough about all things that kids need to know to grow up and become functional, vibrant adult members of society.

Virtually every kind of institution one can point to in a modern, complex industrialized society is a form of economic entity. Reproduction enters into the corporate world only in the vaguest sense that one of the reasons anyone holds down a job is that there may be kids to clothe, feed, and shelter. Not least among the howls of outrage from corporate America as women have sought increasing presence in the boardroom are the complaints over maternity leave and day care. Corporate America expects undivided attention from its employees, and the focus is purely economic.

Human reproduction does, of course, have a direct impact on the economic world. The increasingly worrying spiral between human population expansion and the reactive response to develop more land—thereby destroying more habitat—is the fundamental threat to the human future, both in the immediate present and on down the line in the centuries to come. Karl Marx taught us that members of relatively disadvantaged classes tend to view their children as economic resources—fitting right in with the corporate executive view of these same children as a source of cheap labor. Solving the human population dilemma will depend in large measure on breaking this particular form of connection between reproduction and the economic side of human existence. Interestingly, the key here seems to be the education and economic enabling of women: When women gain entry to traditional economic systems of their cultures, birthrates are known to plummet.

Reproduction and economics interact in complex ways in human society. But it remains true that there is no overt place for reproduction in the economic world. Not so with sex. Psychologists have known for a long time that sexual activity in the economic arena is more about power than simple lust. Sex, of course, is bought and sold, packaged and commercialized. Prostitution is supposedly the world's oldest profession. But sex pervades the economic arena in far more subtle guises. The current flap over sexual harassment is all about sexual advances, innuendo, and exploitation at work—largely if not completely and exclusively men harassing women. It is partly a changing world exposing an old male tendency to chase women for basically reasons of ego: Women fit in with expensive cars and all the other symbols of power.

But sexual harassment is probably more about the direct threat to the economic power women are striving to gain—from a male point of view, at their expense. In this sense, male-female sexual interaction at the office takes on the same basic flavor as any set of prejudices and group animosities, nearly all of which are rooted in a fear of one group coming in and taking what those who oppose them fear they will lose. Nine times out of ten that means jobs.

Some biologists portray rape as a vestigial manifestation of the primordial drive to leave as many copies of one's genes to the next generation as a guy possibly can. But anyone who has looked at the phenomenon of rape in any detail will tell you that reproduction has nothing to do with it—even if victims wind up pregnant. Sexual pleasure seems to have very little to do with it. Rape is an angry, violent crime; that anger is about power—or, more accurately, a sense of lacking power. Sex here is a weapon, one that profoundly affects the lives—the very existence, the economic side of life in the most general sense—of both rapist and victim.

And perhaps the very best illustration that sex, reproduction, and the economic sides of life are separate but intricately interconnected is the very existence of homosexuality. Homosexuality confounds both church and biology—both of which proclaim that the purpose of sex is reproduction, for humans as well as mountain lions. Homosexuality may indeed just be about sex. But more often it is about something else as well: the economic lives of the partners. Homosexuals have formed close, loving relationships for centuries. Such arrangements closely resemble heterosexual pair bonds—out of which, of course, all homosexuals emerge.

Now, the scenario I have followed about the emergence of stable heterosexual pair bonds had as its payoff the increasingly protracted period necessary for the physical, psychological, and encultural (cognitive) development of children: a distinctly reproductive payoff. Homosexuals are an enigma to biologists who insist that life is all about passing one's genes on to the next generation. But if nuclear families, meaning at base male-female stable pair bonds (however transient such stability may in fact be), are in part about reproduction, they are also about sex and economics. *That's* what there is in homosexual pair bonding: emotional attachment, expressed in part through sexual activity, *and* economic cooperation, which is bound to involve some form of division of labor. No one, regardless of sex, is equally good as another at all things.

Homosexuality highlights the simple fact that life is by no means "all about" making babies. It reflects the simple yet profound truth that reproduction is the one physiological function not required for any organism's—including any human's—survival. One can get by, as well, without sex, though few if any have really managed to do so.

Sex bridges the world of reproduction and economics in so many ways in human life. It is endlessly fascinating to one and all—so much of the gossip of this world is about who is doing what to whom sexually. Sex contributes perhaps more than any other single factor in our human obsession with ourselves. We spend most of our waking (and all of our dreaming) time contemplating details of human life— our own, of course, but also other's lives as we see them impinge on our own. Sex plays a huge role in this collective self-absorption. TV shows, movies, books, stories in the oral tradition, songs—virtually all are about people (mostly their sex lives). There is very little time left over to consider the nonhuman world. We are too busy just surviving in a bustling world of other humans, too busy trading that $1 trillion a day with each other.

Sex symbolizes, facilitates, and perhaps even lies at the very heart of this monumental self-absorption. It may indeed make the world go round. And though there is nothing wrong per se with our endless fascination with ourselves, we should realize that self-absorption goes right along with a deep sense of self-sufficiency—a feeling, even a conviction, that we have forever escaped the confines of the physical world around us. We *have* abandoned our old position as cogs in the local ecosystems around the world. But we haven't yet awakened to the fact that we face a major reckoning with the rest of the planet: the species, ecosystems, waters, gases, rocks, and soils of the earth. Self-absorbed narcissism is partly to blame. Sex decoupled from reproduction, permeating all aspects of human life, is partly responsible. Sex complicates life, but it also makes it a lot more fun.

LIFE OUTSIDE THE LOCAL ECOSYSTEM

Consider any standard college economics text. It is all about exchange among humans: gross national product, the national debt, production and exchange of goods and services, supply and demand, stock market fluctuations, the relation between capital and labor, depressions, and so on. That's what economics *is*—to human life, that is. Hidden deeply within, of course, are the very same basic issues faced by all living organisms: obtaining food and protection from the hazards of the elements and other organisms. As a truly distinctive hallmark of this change in the economic landscape on which our species has embarked, the organism we have most reason to fear—apart from those that transmit deadly diseases—is us, that is, other humans.

We need food, but few of us produce it ourselves. Far more people practice subsistence farming in the emerging, third-world nations than in the more highly industrialized societies—where family owned farms continue to fail, are sold for housing developments and malls, or are swallowed up by big agribusiness. Hunting for daily subsistence is far more rare, at best an occasional supplement to the dinner pot in all but a very few hunting-gathering societies which manage to hang on to some vestige of the older ways.

Hunting-gathering societies that are integrated into their local ecosystems have all but disappeared from the modern world. Historical and anthropological accounts of such societies reveal a mode of economic life like that from which we have all sprung. When members of Mbuti hunting parties call out to "Mother Forest," "Father Forest," they are affirming their intention to take only what they need.

I take the rather lyrical call of the Mbuti hunting party to represent a conscious acknowledgment of the relation that a human population has with a (more or less) pristine African rain forest ecosystem within which they live. The Mbuti's own words suggest that they see themselves as a people, a population of humans, that play a role in, and are a dynamic part of, that ecosystem.

And that is remarkable, for here is a human population that not only acknowledges that it is part of nature but actually explicitly perceives what the role is. I cannot believe the Mbuti are alone among (now largely vanished) hunting-gathering *Homo sapiens* in holding this view of themselves in nature. More likely, the Mbuti view of their position within nature is typical of all human beings from our inception as a species more than 100,000 years ago up to about 10,000 years ago, when agriculture transformed human existence.

Hunting-gathering societies play definitive roles in local ecosystems. Their economic life is thoroughly culture-dependent. There is generally a clear division of labor, typically with males hunting and females foraging for edible vegetation and doing virtually all the rest of the economic work around the home fires. Hunter-gatherers have traded with neighbors along extensive trade routes for tens of thousands of years. Long before the advent of agriculture, for example, flint from the chalk cliffs of western Europe (notably modern France, England, Holland, and Denmark) was highly prized and found its way to societies far to the east. Humans, unlike the rest of the world's 30 million species, have incorporated a culture-centered tendency for economic interaction—within local systems and between social systems. Social insects may have complex economic relations within a hive, as when a worker returns and signals fellow hive members the directions to a new nectar source. Such economic interactions, however, do not extend to other hives and in any case are not transmitted through anything remotely resembling human culture.

Now think of the strong contrast between an Mbuti hunting party and daily life virtually anywhere in the modern world. In the morning, a ten-year-old Egyptian boy walks the family zebu cow and water buffalo, together with perhaps one sheep and a goat, out to the small plot his family tills—and back home as evening falls. This utterly rural, subsistence-level existence seems at first glance pretty basic, maybe even fairly close to nature. But that water buffalo was originally an Asian species, domesticated and brought over to Egypt primarily as a source of milk and a tiller of lands. Zebu cattle, sheep, and goats are, of course, wholly the products of the domestication process that began in the dim mists of the Agricultural Revolution some 8,000 years ago.

The family does not own the plot of land that it tills; some absentee landlord does. And what is this notion of "ownership" anyway? The Mbuti, so far as I am aware, never had such a concept. They feel themselves a part of the Ituri ecosystem, but *own* it? The concept is surely meaningless in the context of Mbuti life.

Not so in rural Egypt—or virtually anywhere else in the modern world. Nothing symbolizes our step away from, and out of, local ecosystems more than the idea that humans "own" land—the soil, its mineral wealth, the trees that grow on it. Everything. I am fortunate enough to "own" land, and think of it in precisely these terms. Yet it did strike me as funny when a friend, complaining of the high densities of deer mice boldly running about her country cottage, said that the mice "think they own the place." No they don't. We do. Meaning we *think* we do.

By such fiat we "own" the world. We buy and sell, trade, and expropriate land. The water that the little Egyptian lad muscles up with the shadouf (or, more likely now, his gasoline pump) is not from the Nile but from one of the myriad canals cut into the alluvium—canals whose genesis stems from the earliest pharaohs, who controlled the availability of water and thus the land. Political control—unification of Upper and Lower Egypt—stemmed in no small measure from control of the waterways. The Aswan High (and earlier Low) Dam is just the latest in a 3,000-year-long history of Egyptian waterworks management.

Land "ownership" epitomizes the fundamental change toward nature and toward local ecosystems wrought by that first Agricultural Revolution. But it was just the beginning, the trigger of an explosive diversification in economic roles played out by humans in settled, agriculturally based societies. And at the center of this great revolution in human existence lies the issue of population. Enter, once again, the Reverend Thomas Malthus.

Hunting-gathering groups are always small, on the order of ten to forty people. The reason is simple: As we have seen, populations integrated into local ecosystems are limited by the resources available to them. Gorillas living on the slopes of Rwanda's volcanoes dine on bamboo shoots and other leafy greens. Each troop moves on after a day's foraging to a new nest site. They have to—gorillas will eat themselves out of a local food supply very quickly if they don't move along and allow the vegetation to recover. Consciously or not, natural populations practice "sustainable development" of their resources, meaning their food supply, their source of energy. Evolutionists tend to look at it the other way around: Populations are "resource limited"—expanding when times are good and cutting back when food supplies fall short.

But it is at the same time true that no population of plant or animal ever truly exhausts its supplies, everything being equal. The elephants of Tsavo were an anomaly, driven to destruction through overconsumption by both drought and human curtailment of their normal, viable range. Usually, the ecological pulse is more placid. Viruses, even those that kill their hosts, never end up killing off their entire host species. To do so would be to commit suicide. For the very same reason, mountain lions and other large carnivores never kill off all the jackrabbits, mule deer, or pronghorn antelope within their territories. Thus the much-vaunted "balance of nature": The economic "strategy" (to put it in anthropomorphic terms) of each population in a local ecosystem is to exploit its resources as efficiently as possible—but in so doing not to take too much at any one time, for that would threaten the continued existence of the resources themselves. Never kill the goose that lays the golden egg!

Once again the contrast between modern *Homo sapiens* and the rest of the living world is strikingly apparent. *We* have in fact driven other species to extinction, including many of the large game animals of the late Ice Ages: the mastodons, mammoths, and other large game species. True, there are indeed examples of other species performing the same extirpation role that we have ourselves so recently (in the past 50,000 years or so) begun to play. For example, the brown tree snake, scourge of so many islands of the eastern and central Pacific, has driven several species of birds to extinction on the island of Guam. Species such as the flightless Guam rail were easy pickings for the snake, which was not native to Guam in the first place. It got there during World War II and still gets around to other islands by hitching rides on the wheels and in the cargo holds of airplanes. Such events no doubt happened without the helping hand of our species in the remote past. But, to my knowledge, all the recent examples of nonhuman species wiping out other species happened because humans—inadvertently or not—introduced foreign elements into local ecosystems. Left to their own devices, natural systems are (as the overwhelming rule) in a state of dynamic equilibrium. And no one species is eating itself out of house and home.

The issue at the moment is human population growth, the immediate consequence of settled existence and the linchpin of economic diversification that has progressively turned us into the most inner-directed species in the history of our native planet. Recall the bet between economist Julian Simon and ecologist Paul Ehrlich—won by Simon because Ehrlich failed to factor in quintessentially human cultural machinations: Sure, when supplies dry up, prices go up; but there is the (to some) seemingly inexhaustible supply of human cleverness on which economists like Simon (and all who would be unsullied optimists of the human future) rely. Ways are continually found around what is perennially thought of as only a temporary inconvenience, a short supply in some valued commodity or another. To the extent that the Julian Simons of this world are right, they have succeeded, obviously, in pointing to yet another difference between post-agricultural human life and everything else that lives on the planet: We seem able to wriggle free of resource limitations more readily than any other living species in the history of the planet.

But have we really stepped outside of nature, away from the natural world, so completely as Simon would have it? Are we really no longer subject to the usual rules of resource limitations that dominate the rest of the living world? The exponential growth in human population in the past 10,000 years is a simple, direct response to newly available resources. *Reliable* resources. Some anthropologists in recent years have been fond of pointing out the *non*reliability of agriculture. Certainly episodes of famine have hit very hard from time to time and promise if anything to be more common, more of a scourge, as humanity lurches into the twenty-first century. But there can be no doubt of the overall—and lasting—effect that agriculture has had on human population. The numbers are too stark to deny: Agriculture enabled settled existence, and population skyrocketed. Bouts of famine notwithstanding, agriculture has supplied such a sufficiently reliable increased resource base that population has grown pretty much unchecked since its inception.

Population size is crucial to our future. It also lies at the heart of our inner-directed economic system. Hunting-gathering bands, limited in size by the nature of their subsistence and relation to the natural world, were likewise limited in the number of separate economic tasks parceled out to band members. Settled agricultural communities quickly became another story. Supplies of food, sufficiently large to support population numbers far in excess of usual band sizes of hunter-gatherers, quickly turned out not to require the attention of the entire populace. That was as true 5,000 years ago in Egypt as it is today: Not everyone need be a farmer in an agriculturally based society.

But there *is* an upper limit. Population experts have long known that whatever we do to curb our own population growth, Malthusian considerations themselves will level off human population growth somewhere between 10 and 14 billion people. This will happen simply because, despite our having abandoned the local ecosystem as the locus of our economic lives, we have not thereby left the natural world. We have merely redefined our place within it: We interact in a concerted fashion with the global natural economy, which willy-nilly will ultimately set limits on us. The question is: If we wait and let nature take its course, will the results be so catastrophic—in terms of social upheaval, resource depletion, and environmental degradation—that we will not be sustained?

HOMO SAPIENS AND THE GLOBAL SYSTEM

Well enough. We see that we are divorced from local ecosystems. We are not like all those other species that are broken into small units, each integrated into a local ecosystem. And unlike all other species, we focus both our economic as well as our reproductive needs and concerns largely within ourselves—within our own species. We are incredibly self-absorbed, and sex, bridging the otherwise separate worlds of economics and reproduction, only adds to our internal cohesion and self-absorption.

A squirrel spends far more time contemplating members of other species—the trees it feeds on and takes shelter from, the hawks and cats that might eat it, the birds and rodents that may compete for its food—than any human in this postagricultural world ever does. We think almost exclusively of ourselves and each other. We are the only species that *as a cohesive whole* plays a concerted role within a natural economic system. Parts of other species play roles in local ecosystems. As a cohesive species, we play a direct role in the global ecosystem. What does this really mean?

What, we ask first, is this global system? James Lovelock's notion of "Gaia," in its extreme form, sees the earth as a living organism: the crust acting as its skeleton, its surface waters and gases interacting with the thin rind of living creatures that are the system's tissues and organs. The metaphor is fanciful, and flawed. As microbiologist Lynn Margulis—otherwise a great champion of Gaia—points out, no true organism is a self-contained system. No organism survives by eating its own effluvia. But earth (call it Gaia if you will—*Gaia* is simply the old Greek word for "earth" anyway) is indeed a dynamic system with complex reactions going on between its biological systems—the biosphere—and the hydrosphere (its surface waters), atmosphere, and lithosphere (its crustal and lower rocks).

What, then, is the exact nature of our relation with the global system? I have been saying that, with our profound inner-directedness, our 5.7-billion population, our $1-trillion daily international interchange, we have become simultaneously global *and* a truly unified economic force interacting directly with the planet. How does this work?

It is clearly a two-way street. We depend directly on this system for ecosystem services: As visitors from the outside we continue to invade large ecosystems, mining them for their resources. The Bering Sea, one of the most productive marine regions on earth, captures this dual relation perfectly. On the one hand, native peoples, such as the Aleuts of the Aleutian and Pribilof Islands, continue their sustainable harvesting practices of the fish, marine mammals, and birds of their local system. What makes the Bering Sea ecosystem a true part of the global picture is not so much the fact that local peoples rely as well on the world economy—for additional foodstuffs, as well as other products and services, in exchange for services to the international fishing industry. What makes the story truly global is the international fishing industry itself. The world population has to eat; we *need* these resources. But look what is happening to the Bering Sea and many of the world's other rich fisheries—the international fleet is rapidly depleting the stock to dangerously low levels. You can see the effects of mindless overharvesting even on the shores of the islands, where seabird and seal populations—all dependent on fish—are plummeting.

Here is another example: Botswana, just north of South Africa, is generally considered to offer one of the very best hopes for survival of African wildlife. With a stable government and sound economic base, conditions seem just right for successful coexistence between human life and natural ecosystems. But not all is well in Botswana. The huge herds of wildebeests and other antelopes have declined abruptly in the past several decades—in direct relation to the expansion of cattle herds and the proliferation of fences intended to curb hoof-and-mouth disease, fences which have had the unintended side effect of separating migrating wildlife from crucial water supplies during the dry season. The cattle, and the goats sent in after the cattle have chomped over the wild grasslands of the Kalahari, destroy the grassland utterly, and true deserts have rapidly replaced what was once productive land. Desertification compounds human poverty, and far more of the human populace of modern Botswana relies on more food relief than was true but a few years ago. All this is driven by international development concerns eager to promote cattle raising in African nations.

We continue to rely on local ecosystems; but because the vast majority of us are no longer functional *parts* of them, we simply do not see the importance of sustaining them. We have been getting away with this for 10,000 years. But now that we are truly global, we must see that Lovelock's Gaia is nothing more—or less—than the sum totality of all these local ecosystems. How do we as a global, inner-directed species affect Gaia? Up to now, we have been ignoring, exploiting, or trashing just the local systems where each of us lives. The effect on Gaia has been purely additive. The situation has recently grown much worse, as modern industrialized nations reach out to exploit the resources in systems where we ourselves do not live. How does Gaia affect us? Its fragile biological component—the organisms living in all those ecosystems—continues to feed us, and to provide critical medical compounds, not to mention the oxygen, nitrogen, and sulfur and countless other products essential for human existence. We need Gaia. That's why the future looks so worrisome.

6

Fashioning the Future

George Santayana's famous dictum—that those who cannot re-
member the past are condemned to repeat it—has been done to death.
It is a proscription for the future, one that is generally considered wise,
if a bit vague. Because past trends in human ecological history impart a
sense of where we might be headed, Santayana's aphorism seems apt.
Our history tells us how we have come to our present state. And it is
who we are right now—the very fabric of the human condition and
our relation to the natural world—that will determine what happens
next.

But our ecological and evolutionary history has taken us to a state
very different from any condition we (or any other living species, for
that matter) has ever experienced. Here we sit, divorced from local
ecosystems, a global species 5.7 billion strong and growing, a species
that exerts its economic influence on the global ecosystem. This situation is so new, so utterly unprecedented, that there is no way we can
simply check our past ecological situation and make glib predictions.

Nor—to get back to Santayana—can we reasonably expect to repeat any segment of our ecological (or evolutionary, for that matter) histories, whether we pay any heed to that history or not.

It's simply not in the cards. Human ecological and evolutionary states were so utterly different from our present condition that there is no way we will see a repeat of any of the patterns prior to 10,000 years ago. I can see only one realistic possible exception: a variant version of the *On the Beach* scenario. In Nevil Shute's 1957 novel, the human species self-destructs in a nuclear self-immolation. At first there was some hope that the southern hemisphere would escape the ravages of this northern devastating folly. But after a few years, the radioactive cloud drifts over the equator, and the southern hemisphere is doomed as well.

Paleontologist-artist Dougal Dixon's *After Man* paints a whimsical ecological and evolutionary picture of the earth and its inhabitants 50 million years after the extinction of *Homo sapiens*. Weird creatures like "sand sharks" populate Dixon's future world.

Because the continents are slowly drifting around at about one centimeter a year, it is possible to project their courses and the distances they will travel over the next 50 million years—arriving at a presumptively accurate picture of the basic geography of those remote times in the future. Knowing where the continents will most likely be yields some chancier propositions on future climates. Dixon mixes this all together with an unflagging faith in the power of natural selection to fashion an array of utterly novel creatures inhabiting the lakes, deserts, and forests 50 million years hence.

But this all involves a future without us, a scenario in which humans don't figure at all. More than likely, we *will* figure into the scenarios of the future, though surviving for 50 million years—in whatever form— is a formidable proposition. Most terrestrial species endure for no more than 1 or 2 million years.

But Dixon is right in imagining a world populated by many new species—should *Homo sapiens* actually become extinct sometime in the near future. Past mass extinctions, occurring long before the appearance of ancestral human species some 4 or 5 million years ago, bear witness to the extraordinary regenerative powers of life on earth. Ecologically, habitats laid waste by volcanoes—or oil spills—are quickly occupied by a few hardy species who come in as pioneers, setting up the beginnings of new communities. Sooner rather than later, the new communities begin to resemble the old ones living there before the disaster—as the old species come back one by one from surrounding territories.

But mass extinctions ratchet the process up a notch: Whole species become extinct, no longer available to stock rebuilding ecosystems. That's where evolution comes in. In the case of the truly devastating mass extinctions of the remote geological past, new species evolve from the few survivors—and it is out of these that entire new ecosystems are fashioned. As we saw, in intermediate cases where ecological disruption is regional—as in the change from wet woodland to dry savannah in eastern Africa 2.7–2.5 million years ago—new ecosystems are fashioned from a combination of recruits from elsewhere and the evolution of truly new species.

We are now faced with a mass extinction event of somewhat different dimensions. There are many parallels between the dramatic loss of species in the world's ecosystems going on at this very moment, and the major, global mass extinctions of the geological past. Instead of climate—or the devastation wrought by meteoritic impact (the favored scenario for the mass extinction 65 million years ago)—the loss of species going on around us today is caused directly by human beings. The *root* cause remains the same: Destruction of habitat is the underlying cause of nearly all extinction events. We are wiping out so many species (up to 27,000 a year, according to E. O. Wilson) *not* because we are madly racing around, singling species out for extirpation, driving them to extinction one by one.

No, we are causing the present-day mass extinction—the "biodiversity crisis"—because we are systematically converting terrestrial habitats to our own use. We are destroying habitats, not so much for fun (though all-terrain vehicles leap to mind) as for simple profit. There are tremendous ramifications of the wholesale conversion (and sidestream despoliation through pollution) for our own future.

But, for the moment, consider the problem from the vantage point of the rest of the world's species: After every mass extinction in the geological past, life rebounded. Even when perhaps as many as 96 percent of the world's species became extinct 245 million years ago, the world's ecosystems were eventually refashioned, Lazaruslike, from just those few (4 percent) remaining species. But that sort of recovery is simply not in the cards for today's ecosystems—not, that is, as long as *Homo sapiens* is around continuing in its present course.

The big difference from the old climate- or meteor-caused mass extinctions and today's ongoing bout of mass extinction is, as ecologist Michael Soulé reminds us, that the irritant—the source of all that habitat destruction—remains on the scene. If a comet or asteroid strikes the earth, all hell breaks loose, devastation is immediate, but then the dust quite literally settles and a semblance of normalcy returns—how quickly will depend on the extent of the loss of species. Climate change, of course, doesn't happen quite so fast, though abrupt changes in global temperature in the last million years are now being uncovered in the geological record. But the climate-induced ecosystem turnover Vrba documents in her African ecosystems now seems to have taken some 200,000 years to have had its full effect.

The point is that whatever the physical cause, its effects eventually wear off—and evolution takes its course. Someone recently phoned me to ask, Since evolution historically replaces species lost in mass extinction events, and because we are losing species at a rapid clip right now, where are the 27,000 *new* species that must be appearing every year? My answer was only part of the story. I told my caller that extinction events can happen quite rapidly. If a species of beetle is confined to a few hundred hectares of Amazonian rain forest, and those hectares are cut down in one season's cutting time, that beetle species is gone—in a matter of months. Speciation, on the other hand, while perhaps not demanding the huge amounts of time for the old Darwinian picture of slow, gradual change to come into play, nonetheless requires time on the order of thousands of years. In recent years, many paleobiologists have been using the range of 5,000–50,000 years as a generalization of the usual amount of time that speciation must take. It simply takes more time for evolution than extinction.

But that really is not the whole—or perhaps even the most important part—of the story. For there is no dust settling, nor indeed any temperature equilibration built into the story of human habitat disruption. True, cutting forests for timber, and for the conversion of woodlands to stretches of arable land, creates temporary "islands" of habitat—chunks of old habitat now separated by varying distances of clearing. And isolated patches of habitat are important ingredients in the speciation process: The initial fragmentation of an ancestral species may eventually lead to two (or even more) reproductively isolated daughter species. But it does take time for reproductive isolation to be developed, and meanwhile we humans keep chopping away. Those little habitat islands are far too ephemeral to have any significance whatsoever to the evolutionary process.

Unlike climate change, human-caused habitat alteration proceeds too rapidly to trigger the evolutionary process. Our destruction of habitat has purely ecological consequences: elimination of livable habitat for most native species, and creation of habitat suitable for a few others. Few ornithologists in the early days of North American natural history ever got to see a chestnut-sided warbler. Today, they are all over the place (though they have begun to decline, along with so many other songbirds, in just the past ten years). The reason for their great increase over the earliest days of American ornithology: Chestnut-sided warblers love to nest in woodland edges, adjacent to clear patches of grassland. They love, for example, to nest along blazes cut for power lines.

That's the big difference between the human-dominated ecological present and the mass extinctions of the geological past. Because we figure to stay around, to be the irritant that does not go away, we will have purely ecological effects. We will continue to dislodge species; we will continue to drive them to extinction. But because we are still here, and particularly if we continue to do what we have, for the past 10,000 years especially, been doing to the natural world, evolution will not be triggered. The long-term evolutionary restitution of the natural world must await our own demise. And that is something we don't want to have happen.

That, from the perspective of the world's ecosystems—of all of the world's *other* species—is the impact and significance of the present human stance toward the natural world. A major—probably *the* major—issue insofar as our own future is concerned is: Does it make any real difference for us? Does the loss of all those millions of species out there (out where, in a functional sense, we no longer live) make any difference whatsoever to our own future? That's the issue, the fundamental question we must face as we chart a course for our own future. And that's the way we have to tackle our own approach to thinking about the future: alternate scenarios depending on whether or not we temper our present course.

But first there is that lingering alternative version of the *On the Beach* scenario. Let's say a presumably compatible and fertile pair on that Australian beach did manage to make it through, founding a small colony of survivors alongside a few other pockets of humanity scattered around the globe. This is one of science fiction's greatest fantasies—the sense of starting out all over again. The stories differ mainly in the amount of knowledge that is presumed to survive, and the consequent rapidity with which post–Industrial Revolution human society can spring back.

History is replete with stories of fallen empires. Survivors don't like it when their pasts seem more glorious than their present state. Time after time complex city-states fall, and the surviving bloodlines are reduced to a peasant status, often eking out a meager agricultural existence. Forgotten is the fact that there was *always* an agricultural peasantry associated with these societies; the priests, merchants, and artisans are really what disappear in such situations. But these are the very sorts of occupations that mark complex social systems, and those are the sorts of societies where knowledge is gained, transmitted, and added to most readily.

The real issue with these variant *On the Beach*, microsurvivorship scenarios, then, is how far back would the human socioeconomic systems, the very approach to nature, be set? All possibilities have been explored. Different societies might come to the fore if knowledge is retained in some places but not in others, knowledge that would allow a fairly quick rebound to modern technological states. Alternatively, whatever remains of *Homo sapiens* might be reduced to pre–Industrial but post–Agricultural Revolution status—the condition of most peoples surviving the demise of advanced city-states.

The most extreme possibility is perhaps the most interesting. One could imagine a scenario where all knowledge of agriculture is also lost. Human survival would then be predicated purely on the ability of people to hunt and gather. I have already said, at the outset of this book, that I will not dwell on such possibilities in my contemplation of the future. Not that I think such could not happen. But I do think it unlikely that 5.7 billion strong will become so reduced—through war, famine, disease—that all knowledge even of agriculture will be lost. Besides, I prefer to believe (hope, anyway) that we will use our intelligence sufficiently well to ensure a survival, a future, that retains the very best of human accomplishments as long as possible.

Still, *if Homo sapiens* should survive but be thrown back to some form of preagricultural state—if, that is, the basis of our very survival into the future depends on our ability to rejoin local ecosystems—some interesting questions emerge. Would those systems be there for us to rejoin? And would another agricultural revolution, a second stepping away from ecosystems, eventually occur? And if so, would population again rise, and would there be industrialization and all the complexities of life as we know it now? Here is Santayana with a vengeance. It is also the favorite parlor game of every evolutionary biologist.

Well, would those ecosystems be there for us to rejoin? If calamity happened right now, the answer is a qualified yes. Eastern Europe is pretty bleak, environmentally devastated through decades of rampant pollution borne of inefficient and careless industrial and agricultural usage. Other places are not so bad off: The eastern rain forest in Madagascar isn't all gone *yet*, though only a few more years logging there (and in the Pacific Northwest of North America) will see the rest of what remains entirely gone to the ax and chainsaw. The good news is that as soon as humans stop the destruction, nature will immediately leap in and start the rebuilding process.

As to ourselves and whether or not our future would retrace our past trajectory—it depends. Many of us in evolutionary circles these days stress the importance of contingency in evolutionary history, meaning basically that so many things can and do happen to ecosystems and species through the course of time that no way, were the evolutionary clock to be reset, would the results turn out the same.

That is the thesis, for example, of Stephen Jay Gould's *Wonderful Life.* There he maintains that had a different spectrum of creatures survived from the remote Cambrian Period over 500 million years ago, we would have a very different complexion of life with us now—or, rather, on earth right now, for *we* might not be here at all. There is no reason to suppose that if the evolutionary clock were to be reset, anything resembling humans—featherless bipeds with remarkable cognitive capabilities centered on an excruciatingly keen self-awareness—would ultimately, not to say inevitably, evolve. Indeed, there is no reason to suppose that intelligence of the human sort would have appeared in *any* species—dinosaur, insect, bird, or mammal.

But there is the matter of parallelism—the independent acquisition of winged flight in different lineages: flying reptiles (pterosaurs), birds, mammals (bats), not to mention insects. Thus my "depends"—my guarded qualification in answer to the question: Would it all happen again if we were set back (and able to make it) as once again members of local ecosystems?

It depends on how far back we go. Evolution is irreversible—at least in the extreme sense that we, *Homo sapiens,* could ever "revert" to some physical primitive stage. In his novel *Galapagos,* Kurt Vonnegut proposes that we will survive—in fine *On the Beach* style—but only because natural selection will modify our brains, greatly *reducing* our thinking abilities. His moral message is of course that we are too smart for our own good.

More on what selection may have in store for us in a moment. The issue right now is reversion to simpler, long-abandoned styles of existence. If such should ever happen (and I am not predicting that it *will*), it will have to do so wholly within the confines of our species as we exist at the moment. We might turn the clock back in a behavioral, ecological sense, reverting to life as small bands integrated into local ecosystems. But we would have to do that as *Homo sapiens;* there is no going back, in a biological, evolutionary sense. We cannot become, once again, *Homo erectus*—let alone *Homo habilis* or *Australopithecus africanus*.

But once back in those local ecosystems, how likely would it be that we would again discover agriculture? Very likely. After all, we discovered agriculture independently at least five different times in the past—and the true number is undoubtedly much higher. With humans once again living a settled existence, population numbers would be bound to grow once again—and with this, division of labor. Human creative faculties would once again themselves be cultivated. Would the exact form of civilization as we now know it take shape? That, too, depends. We would undoubtedly have an entirely different spectrum of languages, but spectrum there would be. We would have cities, but who is to say where or when wheels would be reinvented, native metals discovered, gunpowder concocted and spread? The details would be different, but I can't help but think that something rather like what we have right now would, sooner or later, emerge.

Time now to turn to the future—or rather the three futures sketched out in chapter 1. The short-term future, hinging as it does on social upheaval arising from unpaid Malthusian debts, is an essentially sociopolitical problem. As promised in chapter 1, as a biologist I will refrain from commenting further on it, except to reiterate the hope that runaway population growth and the disparate distribution of wealth and resource consumption, the main threats to our short-term future, can be satisfactorily addressed. My story here—of who we are and how we came to be this way—is more germane to predicting the intermediate-range, ecological future, and the *only* way we can say anything intelligent about prospects for long-term human evolution. I will take these latter two aspects of the human future in reverse order, preferring to save the critical issue of our ecological future till last.

THE HUMAN EVOLUTIONARY FUTURE

Famous midcentury evolutionary geneticist Theodosius Dobzhansky openly scoffed at anyone who dared suggest that, somehow, human evolution had ground to a halt—now that we had reached some imagined pinnacle of twentieth-century perfection. He had a point. There is a vast array of genetically determined characters in the human makeup, and these are all potentially subject to natural selection, plus the essentially random factor of genetic drift. The frequency of different forms of genes (alleles) is bound to change from generation to generation, as much so in humans as in any other species.

Many examples of natural selection from wild populations—of moths, birds, or humans—involve some sort of balancing action. Sickle-cell anemia in humans is just such a classic case. In some African and circum-Mediterranean populations, where malaria has long been endemic, the frequency of an allele that causes deformation of red blood cells, and reduces their capacity to carry oxygen, is also quite high. When present in double-dose form (inherited from *both* parents), this allele causes full-blown sickle-cell anemia, an often fatal disease. The alternate form of the gene produces normal red blood cells. A double dose of the normal allele yields, of course, normal red blood cells. People with one of each allele display some symptoms of "sickling," but they do not develop full-blown sickle-cell anemia and never succumb to the syndrome.

Now, with such an affliction in the double-dose (homozygous) state, normally the gene would be expected to all but disappear from the population over time—especially if the disease hit before reproductive age. (That's the problem with Huntington's Chorea, a fatal genetic disease that shows up well after the normal reproductive years, thus being routinely passed along before the gene does its deadly business.) But there is a catch: It turns out that people with both alleles—those heterozygous for sickle-cell anemia—have a built-in physiological advantage in combating malaria over those who carry double doses of the same allele. The allele causing sickle-cell conveys an advantage in that situation—so natural selection, instead of simply eliminating the sickle-cell allele, actually ends up keeping it in high frequencies in the population.

There are lots of homilies in this sickle-cell example. First of all, Dobzhansky was surely right: The same basic laws of nature, most especially including natural selection, are still very much applicable to *Homo sapiens.* Second, many examples of selection in action are very like sickle-cell anemia; there is a balance, a trade-off that shifts genetic frequencies around. But often there is nothing that will push for the complete elimination of one genetic form and the ubiquity of another. As we saw (in the case of Britain's peppered moths), genetic evolution often seems to be a matter of dancing around some mean value—just as large-scale evolutionary patterns within species seem to me an oscillation around some basically very stable anatomical configuration.

But sickle-cell raises yet another important point: Eliminate malaria, and the heterozygote no longer has that advantage. Selection would then simply go ahead and eliminate the disease-causing allele from the population. Sickle-cell anemia cannot be eradicated without the elimination of malaria from the environs. And that is yet a fourth point: We may be living functionally outside local ecosystems, but there are still plenty of issues—public health perhaps paramount among them—where conditions in the local ecosystem still very much affect local human populations. And that, of course, is precisely the context in which natural selection works—in local ecosystems.

The human evolutionary future has two distinct aspects. First, what will happen to us, assuming we *do not tinker directly either with our own genetics or with the natural world?* And second, what will happen to us if we *do* tinker with ourselves and the natural world around us?

And of course we will be tinkering. We have been tinkering with the genetics of other species around us for at least 10,000 years. We have been tinkering with our own genes as well—partly as a reflection of our modification of the environment and other species around us, and partly directly. The current swing toward biotechnology and the design and management, for example, of food products and the fight against genetically based disease is simply an intensification of this 10,000-year-old effort. But before we look at this trend more closely, we should ask what will happen to us, in a biological, physical sense, if we simply let nature take its course over, say, the next 250,000 years. Will we change much, and if so, in what way?

The short, quick answer is by now obvious: nothing much. Anatomically modern humans put in an appearance more than 100,000 years ago and haven't changed all that much since then. Yes, we have diversified into various "races," but that diversity is largely skin-deep. We are unlikely to diversify further, simply because the opportunity for geographic isolation, so essential for diversification, is now gone—and will remain gone so long as the human population remains at anything like its present inflated 5.7 billion.

Recall, too, that the norm for species—*all* species—that manage to survive for any considerable length of time is stasis: pronounced stability. Successful species hang on because they continue to recognize and exploit suitable habitat. The old expectation of evolve-or-die in the face of environmental change is gone, superseded by a simple appreciation of a more profound ecological reality: Populations of organisms track changing distributions of habitats far more readily than natural selection can track environmental change by altering the frequencies of genes in populations. "Evolve-or-die" has given way to "move-or-die." And so species tend to have longer life spans than previously imagined. And successful species tend to hang on more or less unchanged.

Homo ergaster/erectus hung on for more than one million years without change. And that particular species-lineage was integrated into local ecosystems, like absolutely all other species on this planet save our own. *Erectus* was less sheltered, more subject to natural selection, than we figure to be today. But even here stability reigns, for each ecosystem presents slightly different conditions, and natural selection acts differently on each local population. No way could the entire *Homo erectus* species have possibly changed in a coherent, unidirection through time.

Well, if that is so of *erectus* and virtually every other species on the planet, what is the effect of our stepping outside and away from local ecosystems? True, disease vectors emanating from the environment— particularly if they intersect genetically based characters of immune systems (or peculiar systems such as in the sickle-cell anemia situation)—impinge on our genetic makeup. But to a great degree, our declaration of independence from local ecosystems was also a de facto stepping away from the clutches of raw natural selection.

There is simply no way an increasingly homogeneous breeding pool of 5.7 billion people can be altered genetically in any truly significant evolutionary sense—in the natural scheme of things, that is. In the usual situation, with each local population following somewhat different evolutionary paths (thus summing up to no net evolutionary change for the species as a whole), there was at least a chance for some genetically based novelty to become established in some local population. Geneticist Sewall Wright long ago pointed out that it was at least conceivable that such a well-established genetic variant could be imagined, under certain conditions, to subsequently spread through an entire species. True, this increasing propensity toward genetic homogeneity (or at least interbreeding) if anything will facilitate the spread of genetic information between formerly genetically rather isolated groups. But, again, lack of true isolation should work strongly against any significant new evolutionary novelty ever getting established in the first place.

What is "significant" evolutionary change, anyway? Loss of the sickle-cell allele would certainly be "significant" to those generations as yet unborn otherwise destined to be subject to its afflictions. By "significant" I am really talking about order-of-magnitude of change—such as the large gains in brain size seen at various times through the course of human evolution. And here the loss of any chance for isolation—for differentiation, let alone true speciation, to occur—is obviously crucial indeed. The story of human evolutionary change is rather like that of any other species in this sense: No real evolutionary change occurred without the appearance of truly new species, species that evolved according to the tried-and-true canons of geographic variation and isolation. Without isolation, there can be no true evolutionary change awaiting *Homo sapiens*—still less of a chance that *Homo sapiens* will produce a novel, different-looking descendant species. Once again, provided our population size remains in the same ballpark (or greater) as it is right now, there is no realistic chance for anything much to happen to us in an evolutionary sense.

The real story about future human evolution hinges on the transfer of our ecological strategy from time-honored biological adaptations to cultural means of approaching, dealing with, exploiting, and insulating ourselves from the physical world. The game is changed, and in that sense Dobzhansky was probably wrong to rail against those who had suggested that substantial human evolution is probably not in the tea leaves. The simple truth is that in stepping away from local ecosystems *and in substituting cultural devices for physiological and anatomical adaptations,* we have unwittingly changed the rules of the evolutionary game.

Natural selection is simply no longer out there each generation constantly refining and honing our physical adaptations to the natural world. We have, to a significant extent, been sprung lose from the hard scrutiny of selection.

As to the fantasies of evolutionary change, recall the immensely popular film *E.T.* There are two aspects of this little space traveler in Steven Spielberg's movie that conform to very common projections about the shape of things to come in human evolution. First, there is the matter of E.T.'s large, bulbous-brained head: Physical anthropologists for more than a century have been confidently predicting that natural selection over, say, the next 250,000 years gradually, inevitably, and inexorably will transform our brains into ever larger, more sophisticated thinking machines. All difficulties of birthing aside, I can only offer in demurral the observation that without the opportunities afforded by geographic isolation, perhaps leading to true speciation, there is absolutely no hope that such change could occur. And there is also the matter that no evidence exists that *increased intelligence* is in fact being selected for—whether by natural selection, cultural selection, or any combination of the two.

The second aspect of E.T. we need to look at is his overall appearance: a great swollen head balanced precariously on a frail, emaciated little body. The other, corollary standard prediction of what we humans will look like in our evolutionary future is precisely that: an expanded brain in a swollen head balanced on a spindly, weak little frame.

The inference is obvious. It is mind over matter taken to the nth degree. Our cultural attainments, brought about by an ever increasing brain, are destined (so goes the fantasy) to make our bodies obsolete. Just like cave-dwelling fish and salamanders, many of whom eventually lose their no-longer-relevant eyes, we are destined to lose our no-longer-needed brawn. What of this scenario?

Use and disuse. A fish species evolves in a cave, and the ancestral eyes are lost. Why? Now useless as organs of sight, eyes are delicate structures that can only become injured. They are a potential threat, demanding resources to form, maintain, and heal—resources the individual fish might readily be imagined to have other good physiological uses for. So natural selection will favor those fish whose eyes are poorly formed; selection gets rid of those eyes.

Can we make a comparable claim for future humans, perhaps living in an environment so controlled, so heavily mechanized, that their bodies become, if anything, physiological luxuries that selection will just *have* to pare down? I doubt it. If the present is the key, it is the very highly mechanized societies that command a superfluity of nutritional resources for just those very persons ensconced at the computer keyboard and otherwise buffeted and cosseted by the mechanized trappings of the present world. The response, if anything, is to gain weight through lack of exercise and overeating. Only if resources were somehow to become very scarce while the mechanized scene continued could the scenario be triggered.

But then there is the by now familiar grand objection to seeing anything whatsoever of substance in our biological evolutionary tea leaves: our huge population numbers. The difference, once again, between a small number of pioneer fish invading and taking up isolated residence in a subterranean cavern and *Homo sapiens*, spread all over the globe, is enormous. Once again, I don't see selection acting on us in the imagined direction—and I don't see us breaking up into semi-isolated little pockets, there to see established the beginnings of a new evolutionary destiny. It ain't gonna happen.

So much for letting nature take its course, that is, starting from our present position already fairly well insulated from the forces of the natural world. But there is an enormous area looming in our biological future. And I do mean *future*, as distinct from *destiny*. For our future is in our own hands—and many people think that means, as well, our *biological* future, now concentrated in our cultural hands via the brave new world of biotechnology. Will gene jockeying be the key to our future evolutionary transformations?

DESIGNER GENES

No doubt about it, this is the Age of the Gene. Determining the structure of DNA and discovering how information is carried, duplicated, and transmitted by the simple base-pair structure of DNA and RNA molecules rank among the most important steps forward in science—indeed, in human experience and understanding—of the second half of the twentieth century.

The knowledge itself is exciting and its own reward. It fits in hand-in-glove fashion with the post–World War II arrival of the information era that permeates the cultural scene today. We are an inner-directed species, and we are in constant, complexly networked communication across the globe with one another. The global dynamics of *Homo sapiens* hinges on this instantaneous communication, this constant exchange of information.

Our enthusiasm for the genetic revolution knows no bounds. Genetic engineering, it is commonly assumed, will get us around the Malthusian dilemma: Entirely new foodstuffs and higher crop yields are the assured fruits when we apply our newly gained knowledge of the structure and function of genes to the problem. Who knows—maybe we will even be in a position to control our own evolutionary destiny, wiping out genetic diseases first, then marching on to shape our own future evolutionary transformation. Or so go some of the flights of imagination wrought by the genetic revolution and the burgeoning field of biotechnology.

Genetic engineering *will* play a role in the human future. We will continue the 10,000-year-old effort to shape our domestic animals and plants to our needs through the enhanced techniques of direct genetic engineering. We will continue to attack diseases through genetic engineering as well. These include infectious diseases—with the development of genetically engineered vaccines (in the offing soon, one hopes, for AIDS). And they also, of course, include genetically inherited human disease, an area where progress is being reported regularly.

The human genome project, currently under way, is a massive undertaking. The goal is to decipher every last little bit of genetic information encoded in the DNA of the twenty-three pairs of human chromosomes. There are an estimated 100,000 genes on our chromosomes, each one composed of a number of base pairs (the coding constituents of DNA). Each gene holds the blueprint for creating a product—generally a protein used either for construction of body parts or as an enzyme mediating the complex array of chemical reactions taking place within each of the body's billions of cells. Sequencing every last bit of our DNA is truly a mammoth undertaking, a fifteen-year effort that will cost several billion dollars.

Supporters of the human genome project, of course, see the potential payoffs far exceeding the time and trouble to map all human genetic information. But there *is* a hitch: A lot of human DNA does not code for anything at all. It is "junk" DNA, often consisting of seemingly endlessly repeated segments of DNA sequences that are never transcribed into a definitive product. Some such sequences appear to be defunct genes that once, in our ancestors, did elaborate useful products. Sections of the genes used to produce the all-important hemoglobin molecules (responsible for carrying oxygen from our lungs throughout our bodies) are no longer transcribed, but they look for all the world very similar to the parts of the genes that *do* code for hemoglobin. Other noncoding sections, particularly those that are represented many times over, are more mysterious in their origin. But one thing is sure: these noncoding segments are doing nothing at all except forming bridgework between actively functioning segments of DNA (itself arguably an important function).

Critics of the human genome project argue that it would save time, trouble, and expense to focus just on those genetic sequences that are known to be active genes—sequences of the genome that actually are transcribed into physiologically active products. Why waste time gathering information that will never, in all likelihood, be used? And there it is, the real rationale for the human genome project in the first place: not so much knowledge for its own sake but knowledge for shaping the future. At the forefront comes the ability to fight genetically based diseases, such as Huntington's Chorea, certain cancers, and so forth. But the human immune system is also under tight genetic control, and eventually pathways for fighting infectious diseases might lie in manipulating directly our own chemical defenses. And beyond these salubrious prospects lies the promise—or specter—of exerting direct control over future changes in all manner of human biological properties: biochemical, physiological, and even anatomical.

Like it or not, human *cultural* control over our own *biological* properties—and those of other species—is at hand. It has actually been with us for some time now—at least since domestication of animal and plant species began some 10,000 years ago. It probably began even earlier, as human hunting techniques improved, pressuring prey species and forcing a change in their behavioral survival strategies. Virtually all animal species in the Galapagos Islands show little fear of humans—all save the most recent arrivals, such as vermilion flycatchers, which brought their wariness with them from mainland South America. Similarly, gray whales, long hunted and thus very shy of humans, have recently begun to lose their fear of us—in areas such as Bahia Magdalena on the Pacific side of Baja California, where they overwinter and are strictly protected.

If we have long been active in exerting a culture-generated control over the genetic futures of other species, the same goes doubly for ourselves. Some degree of cultural control over human genetic destinies also began long ago—whenever the first drugs and palliatives were extracted from plants to cure a malady that, if left untreated, would have killed those afflicted. In recorded times, we can point to a vizier, architect, and savant of the Third Dynasty of Egypt's Old Kingdom called Imhotep, a real person who was deified 2,000 years after his death, celebrated as the founder and patron saint of Egyptian medicine.

When culture interferes with the course of disease, we are tinkering with the genetic future of *Homo sapiens*. The eugenics movement sprang to life in the late nineteenth century, fast on the heels of Darwin's *On the Origin of Species*. Darwin himself was no "social Darwinist," but most eugenicists certainly were. Spearheaded by Darwin's nephew Francis Galton, eugenicists voiced concern that medicine was coddling the masses, contributing in effect to the delinquency of our genetic future in allowing the genetically unfit to survive and reproduce. As a counter-remedy, eugenicists proposed restrictions on breeding privilege of members of families with pronounced heritable disease. And they also proposed various programs of selective breeding, modeled on animal husbandry, where only those with superior, desired traits would be allowed to breed and thus enhance the genetic constitution of *Homo sapiens*. Adolf Hitler had his eugenicist admirers, and I need not belabor the obvious: Desired traits always conform to the characteristics of the people proposing the eugenics programs.

These issues are important. Do eyeglasses and false teeth—to take mild examples with which I am personally intimately familiar—weaken the genetic constitution of the species as a whole? When such devices (not to mention much more serious interventional contraptions and medical procedures routinely used to prolong and enhance quality of life) are used, we are applying culture over biology. We render biology superfluous. Only if we were to revert to some *On the Beach* sort of minimal survivalist state—where the ability to manufacture corrective lenses would be lost—would the incidence of severe childhood myopia become a survival issue. People like me (short-sighted from an early age—odd for a person peering into the future!) would presumably have a much more difficult time just surviving under such conditions. Natural selection would immediately set in, and our collective vision in a reversion-to-hunting-and-gathering condition would soon sharpen up. No worries there.

In other words, medical attention to various anatomical and physiological deficiencies, diseases, and syndromes poses no threat to our evolutionary future simply because culture has overlain natural selection in these areas. It just does not matter, and meanwhile countless lives are saved and enhanced—lives as intrinsically valuable as those lived without the difficulties of poor eyesight, missing teeth, or any of the far more serious heritable diseases. Bad eyesight is not a problem precisely because we can correct it artificially.

Only eugenicists and some modern evolutionary biologists still harbor worries over the supposed negative impact of corrective medicine on the genetic constitution of *Homo sapiens*. The rest of us take it as a matter of course—part of the very fabric of human cultural-biological existence. But there is an issue that still lurks in the shadows of the old eugenics debate. If medical intervention arguably increases the incidence of "bad" genes in the collective human population, can we achieve the reverse by direct manipulation of the human genome? Can we rid ourselves of "bad" genetic information? We have (apparently) isolated the few remaining smallpox viral cultures down to just two laboratories where they are kept under lock and key. (Arguments rage on whether those few remaining stocks should be destroyed. They should be—the only situation in which I am in favor of the direct extirpation of [in this case] a quasi organism, a separate stock of genetic information in the natural world.) Can we isolate and elimi-nate, in like fashion, dangerous, disease-causing elements of our own human genome?

With all the brave-new-world talk about cures for genetic afflic-tions, and all the scary scenarios of further tinkering with the human genome, one simple fact seems to be routinely overlooked: All manip-ulation of genetic information is on an ad hoc, case-by-case basis. We are talking of adjusting the genetics of individuals so that whatever deleterious or missing gene product causing their suffering is amelio-rated. This is very much like corrective lenses overriding myopia, but it's more permanent. With gene therapy, basic biological elements of a person's makeup are permanently tweaked.

But, as we have seen, what happens to an organism's body during the course of its lifetime has nothing to do with the genetic information it passes along to its offspring. That's why all so-called use-and-disuse or Lamarckian theories of evolution fail: Giraffes did not get their long necks through a series of neck stretchings by ancestral individuals, who then passed along their stretched necks to their offspring. The genetic information in each of the cells of your body comes, in equal doses, from your mother's egg and your father's sperm. The fertilized egg divides to form two cells, each of which divide to give four. Then, in rapid succession, eight, sixteen, thirty-two, sixty-four, and so on as the embryo grows, tissues differentiate, and organs form. Each cell inherits an exact duplicate copy of the original set of twenty-three pairs of chromosomes—exact, that is, except when mistakes of copying come in: mutations. Along the way, as organs are formed, the embryonic precursors of ovaries and testes form—the organs whose tissues will soon be producing eggs and sperm, respectively. Nothing that happens to any of the other cells of the body after the ovaries and testes appear will have any effect whatsoever on the nature of the genetic material in the ovaries or testes of any other human being.

The genetic information in sex cells, when united to form a fertilized egg, determines the genetic information of all the body cells of an organism. But the reverse simply is not true: The genetic information housed in an organism's body cells has no effect whatsoever on the genetic information of the sex cells. This means, in a nutshell, that whatever form of gene therapy is devised, it is not transferable. Someone cured of cystic fibrosis by such methods will still harbor the defective genetic information, potentially passing it along to offspring. Only if sex cells themselves are genetically altered—or, as seems more likely, genetic alteration is performed shortly after fertilization—will the artificially altered genetic information get into the sex cells and thus be passed along to succeeding generations.

That day is indeed coming, though as yet it hasn't truly arrived. For the foreseeable future, gene therapy of human illnesses will have precisely the same overall effect on the collective human genome of *Homo sapiens* as all other medical interventions have had to date: Individuals harboring genes for particular ailments will enjoy better prospects for survival and a full functional life. And that of course includes reproduction.

But gene therapy, important as it is, is only part of the picture. Can we, and will we, alter important aspects of our physiques, physiologies, and behavioral capacities and propensities genetically? For we humans discovered the direct way of altering not species but artificially isolated *strains* of natural species simply by the uncomplicated process of selective breeding. In the course of the last thousand years, we have wreaked a prodigious amount of genetic change on a wide spectrum of species—if, that is, the panoply of variation within our domestic breeds is an accurate indication of degree of genetic change. (The correlation between degree of anatomical and genetic change is sometimes not very close.) Just think of all the different breeds of dogs there are, or reflect on the stark fact that cabbage, brussels sprouts, kale, cauliflower, broccoli, and kohlrabi are all the same biological species, *Brassica oleracea*. For all our tinkering, we seldom if ever have created an entirely new species—though the domestic dog species, *Canis familiaris*, which encompasses absolutely all dog breeds, is not recorded living freely in the wild.

Our modification of existing species to suit our purposes is very much a piecemeal operation. True, we have had an enormous net effect on the genetics of domesticated populations of *Brassica oleracea*. But that is not the same thing as taking a single species and altering its genetics in some single, wholesale way, some way in which the entire species becomes transformed. Domestic breeding, in other words, works under the human hand very much in the way evolution works in nature: Natural selection cannot and consequently does not modify the genetics of an entire species across the board. Rather, natural selection works on parts of species—populations integrated into local ecosystems. So too with domestic breeds, which are isolated lines, "populations" often living (unlike natural populations) in various different human-controlled environments (laboratories, experimental plots, kennels, etc.). The breeders determine what is mated or crossed with what; animal breeders, for example, routinely mate males and females from widely scattered regions around the country (these days, around the world). Artificial insemination helps a lot along these lines.

Genetic engineering is just an intensification of this 10,000-year-old scheme—speeding up the old process of selective breeding. The papers daily carry reports of cows with enhanced milk yields, strawberries the size of baseballs, and tomato varieties that keep a lot longer in your refrigerator. With biotechnology, people are in essence inventing mutations—preordained, favorable mutations—and injecting them into agricultural breeds. Pretty neat—and wholly consistent with what we have been doing for a very long time.

But we are not, thereby, transforming entire species as we pursue our intensified path to altering the heritable features of the plants, animals, and drug-producing microbes that are so important to our very survival. And that is a crucial point, as we turn for one last look at possible manipulation of our own evolutionary futures. Yes, soon we'll be able to manipulate fertilized eggs and early embryos of humans with the same ease that we can modify mouse embryos today. The sticky issues here lie in the realm of ethics, rather than technology. Sidestepping those ethical issues, imagining somehow that people will in the future decide (or have it decided for them) to take their evolutionary future into their own hands—to grow, for example, those greatly expanded brains housed in bulbous crania perched precariously on spindly, emaciated frames, in short to transform themselves into something like E.T.—*could* they do it?

No—for the very same reasons that evolution itself is unlikely to forge any great modifications of "our bodies ourselves" so long as we manage to survive on earth. There is just no way biotechnology can transform, in lockstep, 5.7 billion people in any one particular direction. Even if we discovered how the human genome encodes brain size, even if we learned how to alter that information to produce still larger brains, and even if we could produce some altered humans in the lab, it would have to be done the same way that strawberries are made to come in ever more jumbo sizes. It is all done by keeping strains isolated from one another.

Enter, once again, the paleontologist-artist Dougal Dixon, who has produced just such a scenario of human control over our own biological future. Dixon sees the creation of a spectrum of what can only be called "biological castes" of genetically altered people—people who are specialists at a variety of tasks. I was as appalled at this particular exercise of Dixon's futuristic whimsy as I was enthralled by his first book, his *After Man*, where Dixon explores what evolution might do, if the world's surviving species were given 50 million years to recover from the nearly fatal experience of sharing the planet with *Homo sapiens.*

Presumably, "wild-type" regular humans (those of us not tinkered with) would remain—and remain, of course, recognizably the same. But there would be this added human-created human biological diversity—those separate strains, those analogues of dog breeds and the cabbage-cauliflower-kale-and-brussels-sprouts spectrum.

Would we ever do it? Depends, of course, on the will of the despotic control that some people may, in the natural course of things, exert over others. For it would not be done willingly, in a collective sense that it would be a good idea to create separate strains of what, to a "normal" person's eyes, would be monsters. Or so I would imagine. I could, of course, be wrong.

On the Information Superhighway, Technofixes, and the Human Cultural Future

Creativity often seems to spring from the interplay of disparate strands of thought. Science, for example, often leaps ahead when someone gets a bright idea, applying new imagery to a tired old set of observations. Copernicus explained more celestial observations when he suggested that the earth travels around the sun, instead of the other way around. Mathematicians are forever just following out the logic of their systems, without regard for the practical utility their formulations may have. It has happened more than once that physicists have stumbled on some arcane branch of math developed thirty years earlier, finding it just the ticket to analyze previously intractable patterns in real-world, physical data.

That's more or less the way the "information superhighway" strikes most of us cynics. When your local cable TV company suddenly doubles the number of channels coming into your living room, what you *don't* see is a commensurate proliferation of new, quality programming. There is still "nothing on"—all the more obviously so with all those additional channels. Shopping channels are informational only in the technical sense of the term. And they are harbingers of an electronic era when indeed much in the way of shopping, bill paying, and the like will be taken care of simply through a nexus of computer-TV-modem-phone line.

But the feeling is also there that, just like Mother Nature herself, the burgeoning technology of the electronic superhighway abhors a vacuum. Human ingenuity almost surely will rise to the occasion, creating and providing greater substance (at least one hopes) to all those beckoning channels which so far are open drains along the highway.

No question, those elements of traditional culture that require isolation and time to develop have absolutely no chance to change appreciably so long as the human population stays in the billions-of-people range. And if we are to survive at all, it will be in the billions-of-people range. I have already noted that under such circumstances, nothing remotely resembling a new language, tied to a particular cultural region, has the slightest chance of appearing. As Frederick J. Teggart remarked, cultural items like language do tend to drift through time—and we can be confident that usage drift within languages will continue. But our descendants will be speaking fewer, not more, languages as we go along.

Indeed, it is notorious that the electronic media—radio and perhaps especially television—engender a form of cultural homogenization that is particularly acute with language. The standard patois of national network newscasters is distinctly a-regional, neither identifiably "New York" nor "L.A."—nor any place in between. The information superhighway will only exacerbate this situation. Anthropologists are quick to point out that human beings, very much the culprits behind the accelerating extinction of species, are also victims: Societies such as the Yanomami of the Amazonian rain forest, and the Mbuti of the Ituri Forest, face a double form of extinction. They may become physically extinct, or they may survive, but only at the cost of the loss of the vast majority of their cultural heritage. No one speaks the old Hottentot language in southern Africa anymore; all surviving Hottentots—heavily interbred with Malaysian slaves and white landowners—speak Africaans.

Loss of cultural diversity parallels the great loss of species that is gripping the planet right now. Both figure to continue, with potentially disastrous consequences insofar as our own, technologically based societies are concerned, for the foreseeable future. Unless some form of stabilization is achieved. More about that in a moment.

But it is simply not true, as we all intuitively recognize, that *all* forms of human cultural inventiveness will disappear. For new ideas come from individual human brains, and the competitive marketplace is enough to ensure the rapid spread—especially now that we have left the footpaths and are now racing along an electronic superhighway—of any and all clever ideas that come along. Technology has been accelerating, and anyone with half an eye open to look at the pace of introduction of technological novelty ever since the advent of settled, agriculturally based existence would never think to suggest that all of a sudden this spate of inventiveness will come to a screeching halt.

I am as firmly convinced that there will be, if anything, an ever quickening pace of technological innovation permeating the lives of just about every individual member of our species as I am convinced that there *won't* be any appreciable, linear, species-wide evolutionary, biological change accruing as we project the human course into the distant future. Cultural change has for too long been decoupled from biological evolutionary change, and its mechanism of change and inheritance too different from genetic information, for such not to be the case.

But what form will these innovations take? I confess I haven't a clue; for such prognostication, it is better to rely on sci-fi visionaries, the only types who have had a successful track record at anticipating the shape of technological things to come. I am simply certain that innovations there indeed will be. And that many of them will have a lot to do with the nature of human existence, and our relation to the natural world around us, in the centuries to come.

Which gets us to technofixes—band-aid responses to emerging problems arising from existing technology. Supertanker on the rocks, leaking millions of gallons of oil, fouling hundreds of miles of coastline and killing sea-bottom communities? Call out the technofixes. There have been some very effective inventions to meet this newly acquired environmental hazard, including bioengineering bacteria to enhance the petroleum-munching proclivities of some species.

Nor are all technofixes in the category of mitigating short-term disasters. Others, like the search for new energy sources, are in a sense less desperation-driven than opportunistic—though I readily concede that we are all collectively looking over our shoulder, wondering what will be found to replace the soon-to-be-exhausted reserves of petroleum to fuel our cars, trucks, and airplanes. But as Marvin Harris neatly demonstrated in his *Cannibals and Kings*, the invention of new technologies, and especially new energy resources, may have solved short-term difficulties but ended up prompting explosive expansion of human population.

Technofixes, in other words, come in several guises. Sometimes we need to solve a specific, short-term problem, and so far we have usually been clever enough to come up with solutions. The worst difficulty with this sort of technofix is that often our corrective measures cause further problems—rather like the problem of sawing down one leg of the three-legged stool because the top isn't level. For example, attempts to reduce acid rain by curbing amounts of sulfuric acid–forming sulfur dioxide in the atmosphere are partly undone by the concomitant reduction in alkaline particles such as ash that are also released—for example, when fossil fuels are burned. It turns out that the alkaline ash particles help neutralize sulphuric acid!

Other sorts of technofixes, though, have even more profound effects on how we fit in with the natural world. The worst one is staring us right in the face: the spiraling interaction between the need to feed burgeoning numbers of human beings on the one hand, and the population explosion that inevitably ensues when resources are in fact increased. The technofix—increased agricultural production—is itself the very heart of the problem for which we need a real fix.

"They'll think of something" is the near-universal reaction that all of us, at one time or another, have invoked when casting a worried glance forward—whether it be to next week's gasoline supply during a gas crisis or to longer-term threats to a continued comfortable existence. To "think of something" is to invoke the inventiveness improving our efficiency in making a living in bygone ancestral millennia, and providing as well the shorter-term, increasingly necessary quick fixes to unintended consequences of our very mode of existence.

So far, in other words, we have indeed been rather good at "thinking of something." And, no doubt, we will go on thinking up neat things. That's what we're good at, and what has brought us to our present state—a state that has many wonderful things to commend it. But it is a state also marked by out-of-control human population growth and its direct consequences: rapidly escalating degradation of the physical state of the planet's surface, mushrooming destruction of the earth's terrestrial and aquatic ecosystems, loss of thousands of species every year, loss of less technologically advanced human societies—all of which lead to some grim prognostications of our own midterm ecological survival.

We really do need to fine-tune our story about who we are and how we fit into the natural world. For projections of our past and present state into the short- and midterm future are sufficiently frightening to suggest that we need a radical course correction to ensure survival in an acceptable quality-of-life state as we move into the next few centuries. If we don't, there won't *be* a long-term evolutionary and cultural history to worry about.

WHAT WE FACE: THE MIDTERM ECOLOGICAL FUTURE

We have become the world's first and, so far, only inner-directed species—divorced from local ecosystems and with an impressive ability to manage the production of our own food, clothing, and shelter. The illusion of being entirely self-sufficient, in effect no longer needing Mother Nature for anything whatsoever, so strongly conforms to everyday life experiences of such a huge proportion of the 5.7 billion of us currently alive, that it cannot be airily dismissed as mythic. Our sense of dominion, as we have seen, does agree in large measure with the independence declared from ecosystems as agriculture took hold and mounting numbers of people took up settled existence.

On the other hand, complete independence from, and dominion over, the natural world is far from our grasp—as becomes only too clear during times of natural disaster. U.S. President Bill Clinton, for example, remarked in an interview shortly after the strong earthquake that struck the Los Angeles area in early 1994 that "we don't have complete control." He meant control over nature—and not, for example, over federally mediated relief efforts to aid the human victims of the quake. We have done much to "do something about the weather," but that something almost wholly takes the form of defenses—houses with internal climate control and sophisticated clothing. We still can't predict the weather with any real accuracy more than seventy-two hours in advance in most situations. And, rain dances and cloud seeding aside, we can't act to change the weather even if we know what's coming.

We still harvest the sea and are if anything increasingly dependent on petroleum reserves and mineral resources. We cling to the concept of "natural resources"—part and parcel of the Genesis story of hegemony over the entire earth; we still naively tend to see the mineral wealth and the world's species as all put there for our own collective personal use, as "natural resources." From the vantage point of such assumed entitlements from the natural world, we humans prefer to assert our dominion rather than independence.

But natural disasters and dependence on mineral and biological wealth for our continued sustenance aside, we see ourselves—reasonably accurately—as outside the natural stream of things. Environmental awareness, while growing in some sectors within some industrialized nations, is still not the preferred vision—unless, of course, one happens to have bought a house somewhere like Buffalo's "Love Canal" section and thus been exposed directly to the health risks and consequent political hassles that come from such environmental hazards. We simply are too accustomed to taking the air we breathe, the water we drink, the soils in which we grow our vegetables, and the domestic and wild animals that we eat, to be pure—to sustain us, rather than themselves to pose direct threats to our health.

Conservationists are finding it increasingly difficult to convince the uncommitted that ongoing loss of species in the wild is a bad thing. People just don't get it, and once again, it is not difficult to understand why this is so. Most of us live in urbanized regions, far away from relatively undisturbed ecosystems. We don't even see the farms producing our food. We operate as if food is produced in the grocery store, water is generated in the kitchen tap, and heat is made in the furnace. It is the very success of our system that allows us to treat these essential lifelines as "black boxes"; when the economy is humming along and most people are living in households sustained by gainful employment, there would seem to be no immediate functional need to know much about these things. Comedian Jack Paar once showed a "documentary" of Italian peasant women harvesting drooping strands of spaghetti from trees, trundling them back to their village in wheelbarrows, then setting them out on racks to dry. Irate viewers wrote in castigating members of Paar's live audience for cruelly laughing at the agricultural habits of these humble peasants.

Well, if we are outside local ecosystems, and are so removed from basic lifeline operations that we don't even know the most basic details of our own food production, small wonder people have a hard time getting concerned that a particular species of pine tree in the Pacific Northwest, or a particular seabird species, let alone some as yet undiscovered species of beetle in the depths of the Amazonian rain forest, is under imminent threat of extinction. How can the demise of any of those species possibly have an impact on us? The "biodiversity crisis" leaves the great majority of the world's human citizenry deeply unmoved.

Even conservationists, people appalled at the thought of the impending loss of even a single species, tend to shy away from the implications such losses might have for human survival—at least as their prime reasons for espousing a conservationist ethic in the first place. To hinge concern over the extinction of nonhuman species on the implications such extinctions might have for continued human existence strikes them as inherently selfish—and not even the best argument for conservation. There are indeed many reasons for people to embrace a conservationist ethic. They include such disparate motives as the (also selfish) utilitarian benefits as yet untapped in bio-diversity—the search for new cancer drugs, for example. Utilitarian arguments extend the "natural resources" metaphor so prominent in human ecological strategies, even those predating settled existence sustained by agriculture. "Conservation" here fits in nicely with "wise use" of all such forms of resources—living species, mineral wealth, water, atmosphere—and as such has direct implications for the human future.

Other arguments for conserving species include purely aesthetic, humanistic, and moralistic concerns. Most arresting of all these, per-haps, is the claim of Harvard biologist E. O. Wilson that all of us, to some degree or other, have an innate *biophilia*—a love of the living world—that has not yet been completely expunged from our post-agricultural, postindustrialized psyches. There is plenty of anecdotal evidence (such as kids leaving the inner-city slums to delight for the first time in the wonders of woodlands, grassy fields, and swamps) to lend some credence to this suggestion. We are, after all, part of that long skein of life that began at least 3.5 billion years ago. It would indeed be surprising that 3.5 billion years of living inside the natural world (in one guise or another) would be completely expunged in 10,000 short years of agriculturally dependent urbanized existence.

But I continue to be drawn to the simple, yet largely unappreciated, fact that should we really precipitate another bout of mass extinction, we ourselves are more than likely to go down with (space)ship earth. It is our own extinction—or great reduction in numbers and quality of life—that to me makes by far the most compelling argument for trying to halt the wanton destruction of the world's species and ecosystems.

But it is a hard sell, this notion that the midrange human ecological future absolutely hinges on stemming the tide of destruction of the world's species—the "biosphere." Once again, we have reason to think we can get along just fine without all those species living out there where we no longer, in a functional sense, live.

But it is nonetheless true that without those species and ecosystems, our prospects are grim. In chapter 5, we touched on some of the immediate consequences we will face should we continue to overexploit and destroy ecosystems. The midrange future looks even worse if we don't take corrective measures right now. Here's why:

Consider, once again, Gaia, the "global system." The energy that drives the convection cells, currents, and waves of the earth's atmospheres and oceans mostly comes from the sun (earth-sun-moon gravitational effects, plus heat released from radioactive decay deep within the earth, also figure in). All ecosystems but those of the deep-sea hydrothermal vents are likewise powered by the sun. The planet's waters (hydrosphere), atmosphere, surface rocks and soils (lithosphere), and organisms (biosphere) all interact in complex ways. In some instances the effects are regional—as when a tropical depression advances westward from the African coast, turns into a hurricane, and runs up the North American coastline, then veers out to sea and follows the Gulf Stream to slam into the British Isles before it finally dissipates.

Organisms, too, have regional effects. Most of the replenished supply of oxygen—a by-product of photosynthesis—comes from the daily activities of the vast numbers of single-celled microorganisms floating near the surface of the world's oceans. Without them, and true plants on land, atmospheric free oxygen would quickly decline.

But most of the interactions between organisms and the physical world around them are of course more local—located squarely within local ecosystems. On land, plants stabilize the soil from which they derive essential nutrients. In death, plants (and animals) give back organic compounds to the soil. With photosynthesis, plants trap solar energy, forming the base of the food chain and enabling animals simply to exist. Plants also transpire water and carbon dioxide (as well as releasing free oxygen)—all of which is critical to the atmospheric balance of gases.

Now, it is easy enough to say that the world will not miss one more species of rhinoceros or big cat. But we have to be careful here, for even large mammals, members of the "charismatic megafauna," play roles in local ecosystems that could prove crucial to the balance, thus the fate, of that system. Recent experiments have indeed corroborated what has long been suspected: that the more species represented in an ecosystem—the more richly complex that system is—the more stable it tends to be. Lions, hyenas, cheetahs, caracals and servals (the latter two, smaller cats), and wild dogs overlap somewhat in their prey items, but on the whole they play rather different roles in the African ecosystems in which all species may occur.

But let's look away from the big mammals. Let's look at the arthropods—the hard-shelled, jointed-leg cadre that *really* own the earth. Insects are arthropods. So are spiders, mites, ticks, and scorpions. So are horseshoe crabs and all crustaceans, including lobsters, crabs, shrimp, and barnacles. Some are disease vectors (mosquitoes carrying the malaria protozoan); some are pests (some barnacles foul ships); some we eat (lobsters, honey produced by bees); some we use to fertilize agricultural crops. And so forth. The vast majority of arthropods are actually more or less neutral to us insofar as our daily lives are concerned, but many people are simply repelled by all arthropods and think in principle that we could get along without them at all—save the few they see as necessary adjuncts to human life.

But consider what E. O. Wilson has to say about arthropods in his book *Biodiversity* (1992). True, Wilson is an ant expert and a lover of the natural world. Still, his words should make the most ardent arthropod hater sit up and take notice:

So important are insects and other land-dwelling arthropods that if all were to disappear, humanity probably could not last more than a few months. Most of the amphibians, reptiles, birds, and mammals would crash to extinction about the same time. Next would go the bulk of flowering plants and with them the physical structure of most forests and other terrestrial habitats of the world. The land surface would literally rot. As dead vegetation piled up and dried out, closing the channels of the nutrient cycles, other complex forms of vegetation would die off, and with them all but a few remnants of the land vertebrates. The free-living fungi, after enjoying a population explosion of stupendous proportions [fungi derive their nutrients from dead organic matter] would decline precipitously, and most species would perish. The land would return to approximately its condition in early Paleozoic times, covered by mats of recumbent wind-pollinated vegetation, sprinkled with clumps of small trees and bushes here and there, largely devoid of animal life. (p. 133)

Pretty graphic stuff. As just one example of the sort of thing Wilson is getting at, consider termites—so loathed as pests by homeowners the world over. Termites can literally eat you out of house and home, and most people on principle wouldn't care if termites were expunged from the face of the earth. Most people would probably think it would be a great idea. But it wouldn't. Termites eat wood but cannot digest cellulose. Few creatures can—and among these are certain forms of spirochete bacteria (others of which cause diseases like syphilis) and single-celled microscopic protozoans. These microbes live happily in the hindgut of termites—enabling their host to absorb nutrients from cellulose, while obtaining a steady supply of cellulose in the right form for themselves to live on. It is a classic case of symbiosis. And it has an enormous impact on the world's environment. Without cellulose digestion, decay of plant material would be greatly retarded, contributing to the effects Wilson so vividly portrays. Termites, it turns out, are enormously important for recycling a huge portion of the world's biotic materials. No recycling, no ongoing life.

There are countless such stories. The microbial world is largely responsible for maintaining the atmospheric balance of carbon dioxide (CO_2); we animals send off carbon dioxide as purely a waste product and think of it almost as a poison. Without the methane-producing bacteria—source of most of the world's carbon dioxide—atmospheric oxygen would soon rise to levels that would threaten to fuel unstoppable fires. And without nitrogen-fixing and sulfur-releasing bacteria, there would be no way for these two critical components of proteins—the very stuff of our bodies—to enter into biological systems.

Our very connection with the physical world is made possible by these and myriad other living organisms. It is not just a simple question of commandeering the photosynthetic capabilities of a few plant species and forgetting about everything else out there. We still very much need that "else"—all those other kinds of organisms still living out there in the time-honored, primordial way.

Nor is it a matter of picking out for preservation those kinds of organisms we see we still need, letting the others go. And that is because all these species are broken up into populations and integrated into local ecosystems. Particular nitrogen-fixing bacteria associate with particular plants. Particular spirochetes live in the hindguts of particular termite species. The biological component of the global system is not one single, mega-ecosystem, where, for example, a single species, or even a group of species, acts as a "keystone"—vital to the structure of the whole global system. Only we interact with the global system that way.

Rather, the global system is the sum of all those local ecosystems the world over—systems that interact regionally, and ultimately globally. That's why local ecosystems are still so vitally important for us: We depend on the health of the global system, and the global system depends on the combined health of all the local ecosystems.

So, what of our midrange ecological future? It depends. The data are sufficiently compelling to convince most biologists that we are indeed in the midst of another great wave of mass extinction. There is no doubt that the habitat destruction that underlies all such episodes this time is coming directly from the hand of *Homo sapiens*—a process really begun in earnest as we converted local terrestrial ecosystems to monocultures, and a process only exacerbated by industrialization.

And, as we have seen, this wholesale conversion of terrestrial ecosystems (as well as the sidestream pollution of atmosphere, continental waters, and, now, oceanic regimes) is what underlies all this species loss, and concomitant dismantling of local ecosystems at base stems from out-of-control human population growth. True, our numbers are expected to stabilize on their own—when Malthusian limitations kick in as we saturate the global system. But our species operates under a different set of rules than all other species, whose population numbers are controlled at the local level, within ecosystems. There is the very real possibility that our numbers will not stabilize "naturally" until we have wrecked too many of those ecosystems, thereby crossing a Rubicon and permanently impairing the global system's capacity for providing those "ecosystem services" we still so very much depend upon. We have seen why human population growth has increased so dramatically, and the possible solutions that have been proposed to stem the rising tide: Apart from Draconian measures, and beyond the fantasy of upgrading the economies of third-world nations to industrial strength, the only hope seems to lie in the economic enfranchisement of women throughout the world. When matched with open birth control information and means, the education and economic empowerment of women holds the only promise of respite. And therefore of survival.

The world will be an ugly place for humans if nothing is done. Indeed, the chances are great that the problems of the very short-term future—involving purely geopolitical issues among nations and peoples, and devolving from that very same specter of overpopulation—may prove so hard to solve that the midrange, ecological future may prove to be moot. I hope not, and also basically think not.

The key to both the short-term geopolitical and intermediate ecological futures of *Homo sapiens* is stabilization: stabilization primarily of human population—but also, necessarily along with that, stabilization of the destruction of the physical acreage of the world's local ecosystems. It is not too late to stabilize. The future absolutely depends on it. To reach that stabilization, we have to realize who we are and how we have gotten here. If we are going to make it, we need another version of the creation myth, another story about who we are and how we still fit into the world.

POSTSCRIPT: ON FRAMING A NEW STORY

It all comes back to stories—stories we tell ourselves about who we are, where we came from, and how we fit into the natural world, into the natural scheme of things. As far as anyone knows, people have been telling themselves such stories since time immemorial. And those stories have changed, keeping pace with a changing status vis-à-vis the natural world. The Mbuti pygmies of the Ituri Forest tell themselves something radically different from the story of apartness and dominion over nature that is the very opening, brave-new-world statement of Genesis. The Mbuti knew they were a part of the Ituri ecosystem. The

Israelites of the early agricultural systems of the Mideast a few thousand years ago just as clearly saw themselves as outside their local ecosystem: So stark was their vision that they proclaimed their condition not so much a newly achieved state, but rather a reflection of the way God had created things in the first place. No acknowledgment of the past here! To the framers of Genesis it is the primordial, wholly natural state of affairs that humans should be outside the living world, holding power over it, seeing it created for the sole purpose of sustaining human life. Such was the new slant on the living world: It is all about people, a system set up by God for humans to control and to use as they please.

This vision has worked for 10,000 years primarily because there has until recently been enough room and resources to support unchecked population growth. Not until the last 100 years or so have the mutually resonating factors of industrialization, increasingly mechanized and chemically dependent agriculture, and out-of-control population growth come together in a horrible form of synergism to threaten the world's species, the world's ecosystems—and ourselves—virtually forcing us now to rethink how well Genesis and similar stories actually work in describing who we are and how we fit into the scheme of things at A.D. 2000.

We need a new vision, a revised story of who we are and how we fit into the world. We can frame any sort of story we want, of course—and there is absolutely no chance that all of us, even all of us in a single industrialized nation, will come to agree on any one particular story, version of events, or plan for the future. But we have to try.

I have laid out my own take on who we are and how we got this way throughout the course of this narrative. In the next few pages, I boil these points down, concluding my own revised myth of human past, present, and future. Here, then, is a brief reprise of my argument, what

I have been saying about who we are and how we have developed ecologically up to this point:

Part of the story of how different we are from any other living creatures comes from our awareness of our own self-awareness: we are conscious creatures and so by definition aware of own consciousness. As a direct consequence, we are extraordinarily self-absorbed and inner-directed—not just as individuals but as a species as a whole. Our economic and reproductive lives are intertwined in complex ways only hinted at in a few other, manifestly social species. The human condition is vastly more complex because sexual behavior forges entirely new connections between reproduction and economics, and itself adds a degree of self-absorption not even hinted at in any other species on the face of the earth.

The result is that, despite our huge numbers, we are extraordinarily cohesive as a species. We spend most of our time worrying about what is going on with members of our own species elsewhere around the world—absolutely unlike any other species that has ever existed. That $1 trillion *per day* that we exchange among ourselves is a most riveting statistic, one that reveals just how mutually inter-reliant we truly are.

This interdependence leads us to think that we are truly self-sufficient. We control production of most of our food—limiting harvesting of noncultivated species essentially to the world's fisheries and the occasional local hunting of terrestrial species that survive (generally seasonally) in isolated pockets around the globe. It is all of a piece: We have indeed stepped outside local ecosystems; we are indeed largely self-sufficient, relying on an amalgam of "natural resources" (which we think of as ours, put there expressly for our personal use) and agriculturally produced food resources for our continued existence. And we have become literally global—a functional, dynamic interacting mass of humanity that exerts, as a unified entity, a direct, economic impact on the earth's mega-ecosystem as a whole.

No other species, as a whole, has ever been such an economic entity; all other species are broken up into small populations, integrated into local ecosystems. Those populations are economic entities, each having its own "niche" in those local ecosystems. Our ancestors were that way—and, as we have seen, some isolated (and rapidly vanishing) pockets of our own species still cling to this primordial mode of existence, albeit with purely human cultural modes defining their ecological "niche." But most of us live differently, ensconced within complex, agriculture-based, and for the most part industrialized social systems.

Thus the story that we in the Western world have been telling ourselves for the last 10,000 years or so is only partly correct: We have indeed stepped outside of local ecosystems. But we have not stepped outside the natural world. We merely changed our stance toward that world. Until our population began to reach its current huge proportions, our status vis-à-vis the rest of nature was equivocal. But now it has emerged as a full-fledged reorientation: We stand foursquare as interactors with all the dynamic elements of the earth's natural system—its atmosphere, hydrosphere (lakes, rivers, and oceans), its soils and rocks. And its species. We harvest; we alter the physical surroundings—mostly inadvertently as we pursue our inner-directed ways of realizing total control and self-sufficiency.

As we have seen, this current functional state of humanity is a straightforward elaboration of an ecological course we have inherited from our remotest hominid ancestors. We started out—in an early "australopithecine" state barely differentiated from the ancestors of the modern great apes—as ecologically generalized beings (thoroughly ensconced, of course, as local populations within local African terrestrial ecosystems). Our lineage—from *Australopithecus afarensis*, on up through *A. africanus, Homo habilis/rudolfensis, Homo ergaster/ erectus/heidelbergensis*—has remained steadfastly omnivorous, despite occasional forays into specialization, such as the obligate vegetarianism of our collateral kin, the robust australopithecines.

Thus we are—and have long been—ecological generalists, jacks-of-all-trades. If anything, we have become more so as the ages have rolled along. Ecological generalists, of course, are nothing new under the evolutionary sun; horseshoe crabs and cockroaches are also generalists. But, once again, *our* form of ecological generality is completely different from anything the world has ever seen before. For we have built upon our primordially general ecological approach to nature— paradoxically enough—through the elaboration of a stunning evolutionary specialization: increase in brain size, linked as it is in complex (and so far largely unfathomed) ways with the development of cognition and consequent ever-more-elaborate forms of behavior.

It is our behavior that imparts the tremendous flexibility of our lineage in approaching the physical environment—in "doing something about the weather." The real transformation in human evolution has been the abandoning of 3.5 billion years of reliance on natural selection to mold and hone purely biological adaptations to the physical environment. There is no precise point in time where we can be sure we see the emergence of cultural over biological adaptation as the paramount governing factor in the human approach to nature. The best clue to where in time that point might lie is the breaking of the traditional link between extinction and evolution of species and definite climatic events. The australopithecines and even early species of *Homo* seemed to come and go—in an evolutionary sense—with the onset of relatively abrupt climatic pulses: global cold snaps. Thus even after the first appearance of stone tools some 2.5 million years ago, our ancestors were still locked into a close evolutionary response to such signal events. Indeed, stone tools were the products of *Homo habilis*—

a species that seems to have evolved in response to the transformation of the wet woodlands of tropical eastern Africa into a more arid open-savannah-like ecosystem.

Even *Homo ergaster* (forerunner of *Homo erectus*) seems to have evolved in direct response to another global cooling pulse, this one the onset of the first of the four major glaciations that mark the Pleistocene Epoch, beginning some 1.6 million years ago. But that was it—and perhaps we can take the cessation of such lockstep events in human evolution in knee-jerk response to climatic pulses as the time when culture came to dominate the human approach to the natural world. Recall that it was 0.9 million years ago, when the second major Pleistocene glacial advance occurred in North America and Eurasia, that hominids first got out of Africa—migrating *north* into the very teeth of that cold snap. We know they had fire. They must have had a great deal more by way of cultural accoutrements to enable their tropically evolved physiologies to withstand the rigors of periglacial Eurasian terrain!

Fascinating, these parallel developments in human and cultural evolution. There were some direct connections, of course: brain size increased—in distinctive, stepwise fashion—as new hominid species evolved over the past 5 million years. And for a time cultural change was linked to the appearance of new species—starting, as just mentioned, with the simultaneous appearance 2.5 million years ago of *Homo habilis* and the stone tools made by individuals of that species. The tool kit hardly changed as *Homo ergaster/erectus* spread north out of Africa: so-called Acheulian hand axes hardly changed much for the 1.3 million years or so that *Homo ergaster/erectus* occupied the lands of the Old World.

But then cultural change came increasingly to be "decoupled" from the purely evolutionary change seen in the appearance of new species. The divorce between evolutionary and cultural change is most obvious in the history of our own species, *Homo sapiens* itself. As Ian Tattersall points out, there was a great expansion of utilitarian objects beginning about 30,000 years ago—and a concomitant explosive appearance of carving, painting, and musical instruments of less than obviously simple utilitarian design.

And cultural evolution, once decoupled from purely physical evolution of new species, began to accelerate. Stone technologies began replacing one another at an ever faster pace as the Upper Paleolithic wore on. And since the advent of agriculture and settled human existence, with its concomitant proliferation of labor, human material cultural innovation has skyrocketed.

There are striking parallels between patterns of human cultural stability and change, on the one hand, and human biological evolution, on the other. Both tend to show great stability, punctuated by relatively much briefer moments of true innovation. When the two were in lockstep—in the early phases of evolution, back in *Homo habilis* and perhaps even *Homo ergaster/erectus* days—the similarity in patterns is perhaps not so surprising: Each species had its own cultural inventions, along with its particular set of physiological and anatomical adaptations. The cultural accoutrements were outgrowths of the level of cognitive capability reflecting the capacity of each species' brains.

Perhaps. But it is equally obvious that cultural and biological evolution were not destined to remain in lockstep for long. Despite their striking similarities (and early correlations) in historical pattern, the mechanisms of cultural and biological stasis and change are just too different. They simply had to diverge. Biological evolutionary change, if not absolutely restricted to the actual formation of new species through geographic isolation, is nonetheless heavily concentrated in just such events. Without such isolation and splitting, new reproductive communities simply do not appear—and species tend to remain the same as they continue to track familiar habitats as the environment shifts around. That, or they go extinct. Seldom does an entire species become transformed, lock, stock, and barrel, as the ages roll by and environments change. Natural selection is at its most effective when dealing with relatively small or intermediate-sized populations—integrated as they are into local ecosystems, with specific roles and needs. Natural selection simply cannot modify an entire, far-flung species—one with many populations and many individuals integrated into a variety of local ecosystems—in one simple direction. Nature just does not—indeed cannot—work that way.

Natural selection works through the process of genetic heredity. Cultural evolution depends on individual invention but does not need to wait for the generations to change for favorable variants to be passed along to genetic offspring. Cultural transmission is lateral as well as vertical, pays no heed to genetic lineages, and is intrinsically separate from the sorts of biological processes involved with natural selection. This means, of course, that cultural evolution in principle should be faster than human biological evolution. And it means that substantial amounts of cultural change can occur *within* a single species, in stark contrast to the very limited amount of biological evolutionary change typically possible within the lifetime of a single species.

What, then, of those hauntingly similar patterns of stasis and change in both biological and cultural evolution? Each realm, it turns out, depends a great deal on geographic isolation for the appearance of novelties, for the development of separate, differing "traditions" (meaning true cultural traditions, or local adaptations in the biological evolutionary context), and for the preservation and persistence of those differentiated traditions. Recent breaking down of long-standing walls of privacy between neighboring, differentiated sociocultural systems will surely result in loss of both biological and cultural diversity, as people and cultures are either extirpated or homogenized to some degree. As the human population and electronic telecommunications media both continue to expand, the opportunities for isolation—thus development of truly distinct new cultural traditions, let alone biological evolutionary novelties—shrink away. And right there we have the key to understanding long-term human evolutionary and cultural futures.

With all this in mind, I present, in the concluding section, my version of the new story of who we are and how we fit in with the natural world. We will need something like it if there is to be any future worth having.

AFTERWORD

A New Story

The People came from the earth and were linked through the recesses of time with all other creatures. They were kin of the bacteria, the microscopic ones, the fungi, the plants, and the other animals. And they were nearest to the Apes: the orangutans, gorillas, and chimpanzees.

The People, though, became self-aware and through time came to devise artifacts and customs to help them live. In time, they left their ancestral home and spread throughout the world. Everywhere they went they lived in and were dependent upon their natural surroundings, which they acknowledged openly.

Then the People came to control their own food supply. Everything changed. No longer at home in the local natural world, the People now lived in inner-directed settlements. They invented gods and declared dominion over all the natural world: the rivers and seas, the forests, the plains, the deserts, and the swamps.

The People prospered and their knowledge grew. But soon their numbers grew so great that they saw they had not, after all, truly left the natural world. They saw a limit to the natural resources, and to the production of their own food. They came to see that poisons of the waters, soils, and air threatened them. And they understood that the other creatures—their kin—were vital to their own survival.

The People acknowledged their true past and their newfound problems. They decided to use the very same tool that had brought them along so far: their cleverness. They saw that all creatures—including the People—face limitations and depend on the natural world.

The People decided to curb their population numbers. They determined to curtail environmental damage and the loss of other species. They decided to conserve the world's remaining ecosystems. And they embraced sustainable development, matching economic growth to the carrying capacities of their surroundings.

The People lived. And it was very good.

INDEX